# Evolution, Games, and Economic Behaviour

Fernando Vega-Redondo

OXFORD UNIVERSITY PRESS

*Oxford University Press, Great Clarendon Street, Oxford* OX2 6DP

*Oxford New York*
*Athens Auckland Bangkok Bogota Bombay*
*Buenos Aires Calcutta Cape Town Dar es Salaam*
*Delhi Florence Hong Kong Istanbul Karachi*
*Kuala Lumpur Madras Madrid Melbourne*
*Mexico City Nairobi Paris Singapore*
*Taipei Tokyo Toronto Warsaw*
*and associated companies in*
*Berlin Ibadan*

*Oxford is a trade mark of Oxford University Press*

*Published in the United States by*
*Oxford University Press Inc., New York*

*© Fernando Vega-Redondo, 1996*

*First published 1996*
*Reprinted 1997*

*British Library Cataloguing in Publication Data*
*Data available*

*Library of Congress Cataloging in Publication Data*
*Data available*
*ISBN 0 19 877473 7 (Hbk)*
*ISBN 0 19 877472 9 (Pbk)*

*Printed in Great Britain by*
*Bookcraft (Bath) Ltd,*
*Midsomer Norton, Somerset*

*A mis padres*

# Preface

This is a book on Evolutionary Game Theory from an economic viewpoint. Indeed, to mention the obvious, I should more appropriately describe it as reflecting a particular economist's point of view of the discipline. Even though this view is certainly not unbiased, I still hope that it is sufficiently balanced to be useful and informative.

I have tried to provide the reader with a general overview of the different approaches pursued by the literature in recent years. Some of its developments are treated in detail. For those that are not, at least a summary is included which should help the reader place them in the general unfolding of ideas. In any case, what I have certainly *not* attempted to do is to produce an "encyclopaedic account" of the subject. This, in my view, is hardly of much use to someone who wants to get acquainted with any new field of inquiry. Moreover, the recent surge of evolutionary research has led to a rather diffuse body of literature which is still too unsettled to make such an exercise sufficiently fruitful.

Precisely due to this state of (evolutionary) flux, I have thought it important to place recent developments in the wider theoretical and historical perspective provided by the original (biologically motivated) discipline. This much more mature area of research provides a good framework of reference to understand the approach, questions, and tentative answers put forward by economic-oriented evolutionary models.

Despite this useful biological perspective, this book is essentially motivated by the belief that evolutionary theory affords a versatile and powerful tool for the analysis of non-biological contexts as well; especially, that is, those of a *social* or *economic* nature. Of course, the proof of this "pudding" is in the eating. This is why the book provides a substantial number of applications (mostly of an economic character) which may convince the reader of its wide potential.

To end this short preface, I could adhere to academic ritual and express my thanks to the large number of people who have helped and influenced me in the course of the years. Since this list is exceptionally long, I will simply mention those colleagues and students who have directly participated in the present endeavour. They have read parts of the manuscript (even the whole of it), sometimes detecting errors and imprecisions, in other cases simply offering their honest advice. My sincere thanks to Carlos Alós, Ana Begoña Ania, V. Bhaskar, Larry Samuelson, Joel Sobel, and Chu Lei Yang.

I would also like to express my gratitude to the different institutions whose support and hospitality has made the completion of this book a pleasurable task during part of my sabbatical year. Besides my home institution at the Universidad de Alicante, these are the Department of Economics of the University of California at San Diego, the Institute for Economic Development at Boston University, and the Instituto de Análisis Económico (CSIC) at the Universitat Autònoma de Barcelona. Throughout, I have received financial assistance from the CICYT (Project nos. PS90-156 and PB94-1504) and the Sabbatical Program of the Spanish Ministry of Education.

Finally, I would like to turn to the more personal side of these matters where any expression of thanks is always an understatement. To stress their unique role, I want to focus on just three persons whose disinterested and open-ended support was consistently there whenever I needed it. To two of them, certainly irreplaceable (as biology goes), this book is dedicated. To the other one, my friend (and wife) Mireia, *moltes gràcies*.

# Table of Contents

# 1
# Introduction

Before starting to explore in detail the wide field of Evolutionary Theory, let me begin by posing the following two basic questions:

(i)  What is an evolutionary model?

(ii) Why should we be interested in evolutionary models?

Once some simple answers to these questions are sketched, the reader may still feel inclined to press ahead. With this in mind, the present introductory chapter closes with a plan of what comes next.

## 1.1  What is an Evolutionary Model?

Evolution, as a vague general idea, could perhaps be suitably used to refer to any dynamic process. Here, however, it will be understood in the narrower sense espoused by biology and, following its lead, by other social sciences. Essentially, a dynamic model will be termed evolutionary if its laws of motion reflect the following "forces": selection, mutation and inheritance, appropriately interpreted for the context at hand.

Out of these three forces, selection was the first to be well understood in biology. Darwin, in his *The Origin of the Species*, conceived it in a form that has been essentially maintained to the present date. Since then, the phrase "the survival of the fittest" has cropped up here and there in the most diverse scenarios and for the most varied ends. It has become, in a sense, a well-accepted phrase applied in very different lines of discourse: scientific, of course, but often also political or even ethical.

Loosely speaking, selection is understood in biology as the force which, reflecting the conditions of the environment, determines the chances of survival and "reproductive success" of an individual exhibiting a certain type of behaviour. So generally defined, it is tautological to assert that selection will be important in explaining *short-run* behaviour, be it animal, social, or strictly economic. The challenge, however, is twofold: first, to identify a useful (operational) concept to which this force may be related in a particular context; secondly, to understand the significant *long-run* effects produced by the repeated operation of short-run selection forces.

As compared to selection, the other two evolutionary forces, mutation and inheritance, have been much harder to formulate and understand in the history of biological thought. In recent decades, to be sure, biologists have progressed enormously in tackling them. Despite such progress, and due to their inherent complexity, existing models of theoretical biology only incorporate mutation and inheritance in a rather simplistic fashion.

Roughly speaking, inheritance is to be conceived as the force that links patterns of behaviour across consecutive generations. It is, of course, a precondition for selection to be effective that inheritance be able to transfer behaviour across generations (at least some of its traits) in a sufficiently stable manner. Selection and inheritance are, in a sense, two sides of the same coin in any evolutionary model.

As mentioned, evolutionary coins (rare mutants) come with three sides. Mutation, the third side of the coin, also fulfils a crucial role in long-run evolution. It is the force that, by generating new and previously non-existent behaviour, is able to enrich prior patterns of behaviour so that adaptation to the environment is possible at all.

Inspired by their role in biological settings, these three forces have been modelled to play analogous functions in the still quite tentative instances in which evolutionary models have been applied to social contexts. In this new realm of application, the approach must necessarily be much more flexible and open-ended than in biology. Evolutionary forces in social environments are always multifaceted, often context-dependent, and, in any case, ever-changing. Moreover, as suggested above, it is seldom obvious what is the best operational content for these forces in many social and economic scenarios. For example: Is differential profit the best basis for understanding the interfirm dynamics of industrial competition, or is it instead differential growth rates and bankruptcy considerations which provide the best explanation? As in biology (or, for that matter, any other science), this difficult kind of issue can only be settled empirically.

## 1.2   Why Evolutionary Models?

In the real world, essentially every important decision problem is enormously complicated. Indeed, people are often forced to tackle several such problems simultaneously. Thus, given our limited capabilities of attention, computation, association, etc., it is no wonder that we usually resort to certain relatively simple rules that, according to our past experience, have worked out reasonably well.

On the basis of our past "record" of the performance achieved by such rules, we tend to adjust our relative disposition to adopting each one of them. This, one might dare to say, looks as if we were implicitly carrying out a certain process of internal (mental) selection, quite reminiscent of the phenomenon of biological selection described above.

But, superimposed on this internal process, social and economic environments often display an additional level of selection, now closer in spirit and form to biological selection. In markets, for example, only those firms which do relatively well can survive in the long run. Or, within organizations under some sort of pressure, only those members who perform satisfactorily (always in relative terms) will tend to remain part of it. Again, such selection (with its implicit "inheritance" as a twin counterpart) appears to play in these social settings a role altogether analogous to that played in biological contexts.

The complexity of the environment and our consequent imperfect understanding of it always holds the potential for some further improvements in our array of actions, strategies, rules, etc. Sometimes, the discovery of new, profitable alternatives comes about by pure (unintended) serendipity. In other cases, one aims consciously for new courses of action by combining a limited understanding of the problem with some random sample of new choices. Occasionally, a good, really good, action is found. Then it is selected, thus becoming part of the present and future choice set. A beneficial "mutation" has occurred which will surely affect the evolutionary course of future actions.

Of course, this discussion is meant to be only suggestive, a metaphorical basis for motivating the reader. She (or he) should not conclude from the preceding discussion that evolution in social and economic systems is "just as in biology". Indeed, evolutionary models in biology can and will be important sources of tools and inspiration for an evolutionary analysis of social environments. However, as will be apparent as we proceed, much needs to be built upon such foundations to produce an evolutionary discipline which is suitable for social and economic analysis.

## 1.3 The Plan of What Follows

The next chapter deals with basic evolutionary analysis as it originated in, and has been applied to, theoretical biology. First, its theoretical framework is presented. In it, the population is assumed to be very large (formally, a continuum), either interacting through random bilateral matching to play some bilateral game (the most common scenario) or by "playing the field" (joint population-wide interaction). This framework is maintained throughout the first part of the book (Chapters 2 to 4).

Much of the original evolutionary analysis centred on the key notion of Evolutionary Stable Strategy (ESS), the first formal notion of evolutionary equilibrium to be developed in the early 1970s. Once the concept of ESS is defined, discussed, and illustrated by a collection of examples, its (problematic) existence is addressed. Then, this concept is related to some of the most significant notions of equilibrium used in (classical) Game Theory, paying special attention to some of the so-called refinements of Nash equilibrium. In fact, the ESS concept is itself found to be one such refinement, more demanding, for example, than the well-known notion of perfect equilibrium.

The second part of Chapter 2 deals with different variations of the general framework which raise some interesting issues. First, the focus is on the case of asymmetric contexts where not all participants in the interaction play the same role (for example, contexts where there is an asymmetry between those that own a key resource and those that do not). Then, the analysis turns to interactions that take place within a finite population. The important difference between this context and that of an "infinite" population is that, if the population is (modelled as) finite, a single individual may have *significant* effect on the rest. This leads to a modification of the ESS concept which allows for the consideration of so-called (evolutionary) spite. Its implications are illustrated by means of a simple example of oligopolistic competition.

Finally, the chapter ends with an interesting application of the "static" evolutionary approach to the topic of (cheap) communication in games. Building upon implicit ideas of neutral drift (a phenomenon which will explicitly reappear later on in different contexts), evolutionary stability is shown to provide a natural basis for understanding the endogenous rise of cheap but meaningful talk in games. In certain set-ups (e.g. games of common interest) it is able to achieve players' efficient co-ordination.

Chapter 3 turns to dynamics. Here, the key theoretical object is the Replicator Dynamics (RD), a stylized formalization of Darwinian natural selection which follows directly from an identification of *pay-off* and *"fitness"*, i.e. reproductive success. The RD is first motivated in discrete time, subsequently reformulated in a continuous-time set-up. Then, the relationship between the RD and the "static" concept of ESS is explored. This is carried out on two levels: first, by formulating explicitly the merely implicit dynamics underlying the ESS concept; secondly, by establishing a remarkable role for the ESS concept as a sufficient condition for asymptotic stability of the RD.

In parallel with Chapter 2, this chapter also investigates the formal relationship between the equilibrium notions derived from the RD and those of classical Game Theory. For instance, the requirement of asymptotic stability is seen to induce a refinement of perfect (and therefore Nash) equilibrium. As before, examples are discussed which illustrate the different issues involved.

Next, the RD is extended to the general context where players can adopt (any) mixed strategies. Despite its formal complexity (the resulting dynamical system is infinite-dimensional), the analysis reinforces the formerly established link between the ESS concept and dynamic stability. In this more general context, the ESS condition is not only shown to be still a sufficient condition for dynamic stability, but also seen to become necessary in a suitable "average" sense.

Motivated by the stringent conditions generally required for dynamic stability, Chapter 3 turns its attention towards an analysis that (free from the "ballast" of stability requirements) is concerned with the existence of some long-run regularities. Two key ideas in this respect are the similar notions of *permanence* and *survival*. Both are concerned with the question of whether a particular strategy can be certain to remain present (at a significant frequency) in the long run. Necessary and sufficient conditions for this are explored. As it turns out, they are closely linked to the presence of cyclical behaviour in the dynamic system, which in turn ensures the appearance of well-defined "average behaviour" in the long run.

This chapter also includes a brief discussion of the relationship between the RD and its direct predecessor in the biological literature: the dynamic model of population genetics extensively studied in the 1930s by Fisher and others. Conceptually, these models are mirror images of each other. (Whereas the RD abstracts from genetics and focuses on pay-off interdependence, population genetics adopts the converse stand.) Formally, however, their frameworks are very similar since that studied by population genetics may be regarded as a "symmetric RD". Surprisingly, this symmetry is shown to have very strong implications. For example, in sharp contrast with the complex dynamic behaviour typically induced by the RD, population-genetic dynamics is shown to display both long-run convergence and monotonic increase in average fitness.

Chapter 3 closes with two different applications. The first one centres on the prisoner's dilemma as the paradigmatic framework used to study the long-run evolution of co-operation. It represents the first instance where the reader will be exposed to the idea that the introduction of noise into the evolutionary system (in this case "deterministic" mutational noise) may have interesting dynamic implications. As explained below, this "noise-perturbation approach" plays a key role in the analysis carried out in Chapters 5 and 6. The second application is of a more biological nature. It proposes a stylized model of plant pollination, where the twin issues of costly plant reward and possible "free riding" are dynamically analysed in terms of a suitably formulated RD.

Chapter 4 addresses the study of evolutionary processes in social environments. This requires generalizing the previous theoretical framework in several respects. First, interaction must be allowed to take place among different populations in a possibly asymmetric fashion. Second, the set-up has to be *flexible* enough to accommodate the variety of different factors which underlie evolutionary forces in social environments. To illustrate such variety, two separate families of examples are presented, each one of them reflecting quite different types of considerations. The first one is based on imitation, an inherently "social" phenomenon. In contrast, the second one relies on satisficing behaviour, which is assumed to be carried out by individuals in a fully independent fashion. By imposing specific functional restrictions within each of these two set-ups, the Replicator Dynamics is seen to arise as a particular case.

In general, the essential condition required from any evolutionary system is some notion of *monotonicity*, i.e. the idea that those strategies which fare relatively better tend to grow at the expense of those which do relatively worse. Various manifestations of this general idea are postulated in this chapter (growth monotonicity, sign-preservation, gradient monotonicity), each reflecting either a different emphasis or some specific details of the evolutionary system under consideration (for example, whether extinct strategies may endogenously (re)appear). As illustrated by different examples, which of these alternative specifications should be regarded as more appropriate must naturally depend on the particular context of application.

The first analytical task undertaken in Chapter 4 involves the study of the relationship between the dynamic stability of monotonic evolutionary systems and the concept of Nash equilibrium. In this respect, it is first shown that if an equilibrium of a two-population monotonic system is asymptotically stable, it must induce a Nash equilibrium of the underlying game. The reciprocal statement, however, is typically false. In general, only a very special class of Nash equilibria will qualify as asymptotically stable. For the (monotonic) Replicator Dynamics, for example, only Nash equilibria which are *strict* (thus, in particular, involving only pure strategies) can have this property.

Since stability is such a restrictive demand for monotonic evolutionary systems with more than one population, two alternative approaches are pursued. The first one involves contemplating set-based notions of stability. As illustrated by means of an example, this may provide some interesting "cutting power" in certain contexts. The second approach (similar to that explored in Chapter 3) aims at identifying some of the regularities which the system might generate in the long run, even along non-convergent paths.

For general monotonic evolutionary systems one cannot hope to obtain the long-run "average regularities" which are shown to hold for the Replicator Dynamics. However, a natural and important question deals with the extent to which more general evolutionary systems may at least be successful in discarding non-rational behaviour in the long run. The answer to this question is only mixed: whereas strategies that are strictly dominated by pure strategies (even after deletion of other dominated strategies) can be guaranteed to enjoy only vanishing limit weight, general monotonic evolutionary systems need not be so responsive to other weaker concepts of rationality (for example, strict domination by *mixed* strategies).

Two different applications close Chapter 4. The first one is a stylized Ultimatum game with just two possible proposals (high and low), where the two populations of proposers and responders evolve according to the RD. As in one of our former applications (Chapter 3), the theoretical exercise here involves adding (deterministic mutational) noise to the system in order to "test" the relative robustness of its different equilibria. In particular, the main concern is whether some of the (non-perfect) equilibria where proposers share some of the surplus might be dynamically robust, i.e. have a non-degenerate basin of attraction.

The second application proposes a two-tier hierarchic model of cultural evolution where the population is assumed to be divided into a large number of equal-sized groups, and monotonic selection operates at both the intragroup and intergroup levels. In this context, alternative "cultural" mechanisms for group formation, or the severity of inter-group selection, are seen to have a crucial effect on the long-run prospects of the economy. Specifically, they determine whether it will eventually develop an efficient or an inefficient "convention".

Chapter 5 starts the analysis of genuinely stochastic (or "noisy") evolutionary dynamics. There are two main sources of noise typically considered in evolutionary systems. One of them is environmental noise, such as that produced by random matching or stochastically determined exogenous circumstances. The second one is "mutation", which in social applications is usually interpreted as the result of either individual experimentation or population renewal.

In line with much recent stochastic evolutionary literature, the theoretical framework mostly considered in the last two chapters is discrete. Not only is time indexed discretely but the population is assumed finite. Within every time-period, the population is assumed to be randomly matched in pairs (for some given number of rounds) in order to play a certain bilateral game. As time proceeds, (monotonic) selection forces, matching noise (the only environmental noise considered), and mutation, all interact to produce a well-defined long-run behaviour of the system. Since, due to mutation, the process is ergodic, such long-run behaviour is unique, independently of initial conditions.

In this context, the following basic question is asked: What is the long-run behaviour of the system when mutational noise becomes arbitrarily small? As it turns out, the answer to this question is crucially dependent on the assumption made on the other source of noise in the system: environmental (matching) noise.

Suppose that attention is restricted to a simple symmetric co-ordination game with two actions and two corresponding pure-strategy equilibria. Then, if matching noise remains relatively large (possibly falling to zero, but at a lower rate than mutation), the equilibrium selected in the long run is shown to be the efficient one. In the opposite case (relatively small matching noise, induced by a very large number of rounds every period), the risk-dominant equilibrium (perhaps inefficient) is selected instead.

These contrasting conclusions illustrate the subtle effect played by noise (and, most crucially, *different* kinds of noise) in stochastic evolutionary systems. After elaborating on the intuition underlying this state of affairs, different extensions to other kinds of games and frameworks (for example, a continuous-time stochastic version of the Replicator Dynamics) are also briefly discussed.

Next, this chapter turns to the important issue of how quickly the established long-run convergence will be achieved; or, in other words, how long is the long run. In this respect, it turns out that, as for the long-run conclusions themselves, the *expected* rate at which these conclusions materialize is crucially dependent on the relative magnitude of the matching noise. If large, convergence is relatively fast; if small, it will generally be quite slow since it depends on population size (as the population grows, the order associated with the rate of convergence becomes of progressively lower order). This latter statement, however, hinges very critically upon the assumed *global* pattern of interaction. If, as seems natural, agents are assumed to interact most likely with close-by neighbours (i.e. the pattern of interaction displays a local structure) the rate of convergence will also tend to be fast (independent of population size) in the scenario displaying small matching noise.

Chapter 5 concludes with the following application. Consider any given number of Cournot (quantity-setting) oligopolists who, as time proceeds, adjust their actions by mimicking successful behaviour (i.e. choosing that output which happens to induce the highest current profits). Further assume that they occasionally experiment (or "mutate"). Then, the unique long-run outcome of the process is seen to coincide with the Walrasian state (of course, *not* a Nash equilibrium) where all firms behave as if they were maximizing profits, taking the market-clearing price as given. This application illustrates the following points: first, the wide applicability of the evolutionary approach to contexts very different from those with random pairwise matching usually contemplated by most of the theory; second, its potential, as an analytical tool, to go well beyond a mere mechanism of (Nash) equilibrium selection in games.

The main body of the book ends with Chapter 6. In this chapter, the objective is to enrich the previous evolutionary framework by considering agents who may enjoy a higher degree of sophistication than formerly (implicitly) assumed. Specifically, they are assumed to be able to form expectations on the future state of the process, choosing their actions as an optimal reaction to them.

The chapter starts by proposing a general evolutionary framework of expectation formation, which encompasses the different variations and examples subsequently studied. Since some of these involve asymmetric interactions, the previous single-population approach is extended to the consideration of two different populations. In this context, the evolutionary process is formulated to proceed in two complementary dimensions. First, it has players continuously updating their prior expectations, on the basis of realized (observed) history. Second, it postulates that players react optimally to these expectations by adjusting their prior strategies, possibly with some friction (i.e. only with some probability). As before, the core dynamics of the process (strategy adjustment and expectation updating) is perturbed with some small probability of mutation, now operating on both the strategy and the expectation components. The theoretical question posed also remains the same. It involves determining the long-run performance of the process when the mutation-induced perturbation becomes small.

The key differences displayed by the alternative contexts analysed in this chapter hinge upon the following two considerations. First, the nature of the game; specifically, whether it is simultaneous (with all its information sets visited along every possible path of play) or genuinely sequential (with some of its information sets potentially *not* visited). The second crucial issue concerns the particular postulates assumed on the expectation-formation process. In particular, static and dynamic mechanisms of expectation formation are contrasted.

The first and simplest context analysed is one where players are involved in a simple $(2 \times 2)$-co-ordination game and expectations are statically formed. That is, players implicitly assume that the currently observed configuration will remain in place next period (except, of course, for their own possible adjustment). In this case, the equilibrium selected in the long run turns out to be as in one of the scenarios discussed in Chapter 5, namely the risk-dominant equilibrium. When *static* expectations are the driving force of the evolutionary system, considerations related to relative basins of attractions become pre-eminent, just as when no significant matching noise interferes with monotonic (pay-off-based) selection dynamics.

Attention then turns to asymmetric games (still simultaneous, and with a static rule of expectation formation). As an interesting example, the discussion focuses on a simple bargaining scenario between two different populations. In this context, the former considerations involving the basins of attraction of the alternative (now asymmetric) equilibria can be linked to the corresponding degrees of risk-aversion displayed by each population. If these are different, the evolutionary process is seen to select the equilibrium which is more favourable to the less risk-averse population. This provides an evolutionary counterpart of the analogous conclusion derived from the classical analysis of the bargaining problem.

Next, a genuinely dynamic game is considered. Specifically, it is postulated that the simple and symmetric co-ordination game described above is preceded by the possibility that one (and only one) of the populations may guarantee for itself some outside option. This variation turns out to have drastic implications. For, independently of any risk-related considerations, the efficient equilibrium is now selected by the evolutionary process as the unique long-run outcome. Expectations are still being updated in a static manner. However, the key point to note is that, since not all information sets are necessarily visited along every course of play (for example, the co-ordination game is *not* reached if the outside option is chosen) "expectational drift" may set in on the expectations associated with such information sets. That is, even if mutation on these expectations makes them change arbitrarily, selection (i.e. updating) forces may find no basis to "discipline" them. This, in the end, turns out to permit fast and easy transition towards the efficient equilibrium, by an appropriate build-up of "expectational" drift.

A similar kind of drift also plays a crucial role when expectations are allowed to be dynamic. In this case, agents do not necessarily predict the stationarity of the previous situation, but rely on past history to detect some dynamic (i.e. interperiod) regularities. Expectations, in fact, must *not* be conceived in this context as a certain given subjective probability on ensuing play. Instead, they should be viewed as consisting of a whole *array* of such probabilities, one for every possible history which an agent may observe. It is precisely because of this wealth of possible contingencies on which expectations are conditioned that drift may set in again. In general (certainly so if the situation has remained stationary for a long time), most of the a priori possible contingencies will not have been observed for many periods. Thus, expectations conditional on these contingencies may drift, unaffected by an updating mechanism which is sensitive only to observed realizations. In the long run, this is shown to lead to a state of affairs which is labelled "equilibrium volatility"; essentially, it reflects the lack of any specific criterion for selecting among alternative equilibria.

As outlined, Chapters 5 and 6 involve the exploration of alternative specifications for the evolutionary system. Each of these can be conceived as reflecting different behavioural paradigms on the agents' part. In a sense, they embody alternative assumptions on players' sophistication. Players may range from being, for example, mere imitators (Chapter 5) to being capable of forming expectations on the evolution of the process – be they static or dynamic – and react optimally to them (Chapter 6). In this light, it is natural to ask how the analysis might be affected if players of different types are a priori possible, their corresponding frequencies varying endogenously as dictated by relative pay-offs. An analysis of this important issue closes the chapter, leading to the following two conclusions.

First, evolutionary processes only work as effective (and efficient) mechanisms of equilibrium selection when the upper bound on the sophistication of players is not too high. For example, if only imitators and static-expectations optimizers are possible, efficient equilibrium selection always results. However, if dynamic-expectations optimizers are also allowed, equilibrium volatility (as described above) becomes the prevailing state of affairs.

The second conclusion involves the long-run selection of types, not behaviour *per se*. Since alternative behavioural rules turn out to be fully equivalent at stationary points (i.e. they prescribe the same strategy), the conclusion here is, in fact, one of non-selection: all types may coexist in the long run. As before, drift (now on the type dimension) is also found to play a crucial role in the analysis.

The book ends with a short Appendix, which includes some technical material used throughout. It is collected there in order to avoid interference with the general line of discussion.

# 2
# Static Analysis

## 2.1 Theoretical Framework

### 2.1.1 *Basic model*

Consider a given population of a certain "species", which has a corresponding range of behaviour (or *phenotypes*) available. These phenotypes are conceived as a (finite) collection of *strategies* $S = \{s_1, s_2, \ldots, s_m\}$, each individual adopting one of them in the contest for survival with the remaining members of the population.

For the moment, we shall focus on *symmetric situations*, i.e. situations in which every individual has the same set of available strategies and pay-offs reflect only fully symmetric considerations. Section 2.6 will deal with asymmetric contests. These will be viewed, however, as embedded into an appropriately enlarged symmetric *ex ante* game.

Each one of the possible strategies induces an associated pay-off for the individual that adopts it, given the behaviour profile of the rest of the population. In order to formalize the link between behaviour and pay-off, we define the function

$$\pi : S \times \Delta^{m-1} \to \Re, \qquad (2.1)$$

determining the pay-off $\pi(s_i, \nu)$ of any individual adopting strategy $s_i \in S$ when the vector of frequencies with which the population plays each of the $m$ different strategies is $\nu \in \Delta^{m-1}$.

There are two assumptions implicit in the previous formalization of pay-offs.

First, it is implicitly assumed that the population is so large that every given individual has an insignificant weight. (Otherwise, the two arguments of $\pi$ could not be treated independently.) The important case of finite populations will require a specific formulation in Section 2.7.

Second, the formulation described in (2.1) reflects a so-called *population game*. That is, a context where the influence of the population on the pay-off of any given individual is contained in the anonymous description of the frequencies with which each strategy is being played by the population. This will be a convenient assumption maintained throughout.

As presented, the above formulation is restricted to contexts where individuals adopt only pure (i.e. deterministic) choices in $S$. If individuals may adopt mixed (i.e. random) choices, the simplex $\Delta^{m-1}$ representing the probability measures on $S$ must substitute this latter set as the relevant strategy space of an individual. And then, a strategy profile of the population is a probability measure $\lambda \in \Delta(\Delta^{m-1})$ over the simplex $\Delta^{m-1}$ of mixed strategies. We shall call such a profile a (population) *state*.

The assumption that the population is large ("infinite") and that individuals make their choices independently, allows us to generalize the former pay-off structure in a simple fashion. Specifically, given a certain population state $\lambda$, we implicitly resort to the Law of Large Numbers and identify the expected population profile

$$\nu(\lambda) \equiv \int \tilde{\sigma} \; d\lambda(\tilde{\sigma})$$

with the realized profile of pure strategies. Then, abusing previous notation, the expected pay-off obtained by an individual playing a mixed strategy $\sigma \equiv (\sigma_1, \sigma_2, ... \sigma_m)$ is simply defined as:

$$\pi(\sigma, \lambda) \equiv \pi\left(\sigma, \nu(\lambda)\right) \equiv \sum_{i=1}^{m} \sigma_i \; \pi\left(s_i, \nu(\lambda)\right). \tag{2.2}$$

The essential postulate underlying Evolutionary Game Theory is that current pay-offs determine the relative viability of the different strategies, thus affecting the short-run evolution of their corresponding population frequencies. When applied to biological contexts, this would seem to imply that it is the phenotype (the argument of pay-offs) which is the object of inheritance across generations. However, as is well known today, living beings do not actually inherit behaviour: they inherit the genes that (according to the laws of genetics and interaction with the environment) lead to a specific type of behaviour. Evolution is not Lamarckian.[1] That is, behavioural traits (learned or otherwise) are not themselves the object of inheritance.

In theoretical biology, the explicit consideration of the laws of genetics in sufficiently rich (in particular, sexual) contexts is at present one of its important avenues of research. (See, for example, Eshel (1991) for an early consideration of these issues.) For our purposes, however, we shall find it suitable to abstract from this source of complexity, assuming in most of our models that behaviour is both the object of inheritance and the basis for pay-offs. Only a brief digression into genetic-based models will be carried out in Section 3.8 within the framework of traditional population genetics.

---

[1] This was certainly not the view espoused at the time of Darwin. He himself had a Lamarckian notion of biological evolution – see Bowler (1984). Cultural beings can, of course, "inherit" behaviour from the culture to which they belong. However, this alternative notion of "inheritance" is quite different from its biological counterpart.

### 2.1.2    Alternative scenarios: "Playing the Field" or "Pairwise Contests"

The *general* formulation postulated in (2.2) is consistent with a variety of quite different scenarios.

A very common one is labelled *"pairwise contests"*. It refers to a scenario where every round of interaction has each individual randomly paired with another member of the population in order to play a certain bilateral game in normal form (finite and symmetric). Such a game can be formalized through a certain square matrix $A$ of dimension $m$, the number of pure strategies. The entry $a_{ij}$ of $A$ indicates the pay-off to an individual who chooses strategy $s_i$ against the strategy $s_j$ chosen by the rest of the population $(i, j = 1, 2, ..., m)$. Or in terms of the previous notation,

$$a_{ij} = \pi(s_i, \delta^j),$$

where $\delta^i \in \Delta^{m-1}$ denotes the simplex vector whose $j$th component equals 1 (and others equal zero). For a general profile $\nu = (\nu_1, \nu_2, ..., \nu_m) \in \Delta^{m-1}$ the expected pay-off associated with strategy $s_i$ is given by:

$$\pi(s_i, \nu) = \sum_{j=1}^{m} \nu_j \, \pi(s_i, \delta^j).$$

Thus, in a scenario with pairwise contests, the pay-off function displays the property of being linear in the population profile vector. Combined with (2.2), we conclude that the pay-off function $\pi(\sigma, \nu)$ is, in this case, separately linear (i.e. co-linear) in both of its arguments. This is the key feature of the scenario which has been labelled "pairwise contests".

In the alternative case when the pay-off function is not linear with respect to $\nu$, the context is usually labelled *"playing the field"*. This phrase is intended to reflect a situation where, unlike the case where contests are pairwise and random, the interaction within the population is best viewed as conducted *jointly* at the population-wide level. It does not admit, therefore, an *ex ante* linear dependence on the population profile. Examples of both scenarios are presented below.

## 2.2    Evolutionarily Stable Strategy

### 2.2.1    General definition

The concept of Evolutionarily Stable Strategy (ESS) is a central one in evolutionary theory. It is intended to reflect a stationary situation in the evolutionary process. A situation, that is, in which the pattern of behaviour prevailing in the species cannot be invaded by any mutation which is better fit.

**Definition 1** *A (possibly mixed) strategy $\sigma^* \in \Delta^{m-1}$ is said to be an Evolutionarily Stable Strategy if, for any other $\rho \in \Delta^{m-1}$, there exists some $\bar{\varepsilon} > 0$ such that if $0 < \varepsilon \leq \bar{\varepsilon}$,[2]*

$$\pi\left(\sigma^*, (1-\varepsilon)\sigma^* + \varepsilon\rho\right) > \pi\left(\rho, (1-\varepsilon)\sigma^* + \varepsilon\rho\right).$$

Verbally, a strategy $\sigma^*$ is said to be an ESS if, once adopted by the whole population, no mutation $\rho$ adopted by an arbitrarily small fraction of individuals can "invade" (i.e. enter and survive) by getting at least a comparable pay-off. As explained above, the relationship between relative pay-offs and invadability is essentially a matter of definition in biology, where the pay-off is taken to reflect *inclusive fitness* (i.e. the number of successful offspring). Of course, the empirical issue of measuring such inclusive fitness might nevertheless be quite difficult in particular cases.

### 2.2.2    Alternative interpretations of ESS: monomorphic vs. polymorphic populations

As introduced in Definition 1, the concept of ESS essentially calls for what is known as a *monomorphic* interpretation. That is, it aims to capture an evolutionarily stable situation in which all members of the population adopt a common strategy. Of course, if such a strategy is mixed, the profile over pure strategies induced by it will not be concentrated in a single kind of observed behaviour (i.e. a single phenotype). However, even though realized behaviour may well be *ex post* heterogeneous, any natural interpretation of an ESS must still embody the idea that all individuals "inherit" the same mixed strategy, i.e. it must formalize a monomorphic *ex ante* configuration.

The restriction of the ESS concept to such monomorphic configurations is obviously very strong and quite unsatisfactory. In general, we would like to be able to encompass in our analysis the possibility that a genuinely diverse range of behaviour (that is, a *polymorphic* profile of strategies) could arise in an evolutionarily stable fashion. However, to formalize any such state of affairs, it is essential to overcome the purely dichotomous considerations (mutant *vs.* non-mutant) contemplated by the ESS concept. Instead, one needs to allow for the joint (and dynamic) interaction among different strategies which must underlie any stable polymorphic configuration.

In biological settings, the dynamic framework required to analyse such joint interaction is provided by the so-called Replicator Dynamics, the analysis of which is set out in Chapter 3. We shall discover then an unexpectedly central role for the static ESS concept. First (Theorem 3), it will be shown that, even in polymorphic cases, every mixed ESS – interpreted now as a *polymorphic state* in

---

[2] Note that if $(1-\varepsilon)\,\sigma^* \oplus \varepsilon\,\rho$ denotes the two-point distribution (population state) where $(1-\varepsilon)$ of the population plays $\sigma^*$ and $\varepsilon$ plays $\rho$, then $\nu\left((1-\varepsilon)\sigma^* \oplus \varepsilon\rho\right) = (1-\varepsilon)\,\nu(\sigma^*) + \varepsilon\,\nu(\rho)$.

pure strategies – arises as a strong local attractor for the dynamic evolutionary system. Secondly (Theorem 5), the ESS concept will be seen to provide in addition a necessary condition for dynamic stability (at least in a certain "average sense") if no a priori restrictions are imposed on the set of admissible mixed strategies. Thus, in this sense, the ESS concept turns out to play a very useful (instrumental) role in addressing issues of dynamic stability, even in polymorphic contexts.

## 2.3 Examples

This section discusses two different examples which illustrate some of the ideas proposed so far. The first example involves random pairwise contests; the second one displays individuals who are playing the field.

### 2.3.1 *Pairwise contests: the Hawk–Dove game*

Consider a certain ("infinite") population of the same species competing for a scarce resource. Think of this resource as, for example, land or food given in some fixed and indivisible amount. Each contest for the use of this resource is pairwise, i.e. it confronts two, and only two, members of the species, randomly chosen among the given population. Once the contest is decided, the winner is assumed to extract the full value of the resource. Denote this value by $V$, the same for each member of the species. On the other hand, the loser suffers a cost of $C$, also identical for all members of the population.

The contestants may use one of two strategies. We refer to them as "Hawk" and "Dove", denoted by $H$ and $D$ respectively. We think of $H$ as representing an all-out aggressive strategy; $D$, on the other hand, is to be conceived as a passive strategy, retreating in the face of escalation.

Based on the previous considerations, we shall assume that when $H$ meets $D$, the individual playing the former strategy (the "hawk") obtains the resource and, therefore, the whole surplus $V$. The pay-off of $D$ in this case is therefore taken to be zero. If any strategy, $H$ or $D$ meets itself, we shall assume that the chances of getting the resource are divided equally between the contestants. Thus, the *ex ante* expected *gross* gain from participating in the contest for individuals playing the same strategy is $V/2$ for both of them. The *net* gain, though, differs in each of the two cases: both $D$ or both $H$. If two doves meet, there is no escalation to an actual fight and, therefore, there is no cost incurred from the contest. If, instead, two hawks meet, the expected losses are $C/2$, which have to be subtracted from $V/2$ to obtain the expected net gains.

The previous pay-off description is summarized in the following table:

|   | H | D |
|---|---|---|
| H | $\frac{V-C}{2}, \frac{V-C}{2}$ | $V, 0$ |
| D | $0, V$ | $\frac{V}{2}, \frac{V}{2}$ |

Table 1

Consider now a certain population which adopts $H$ and $D$ in some given proportions, $p$ and $(1-p)$ respectively. For the moment we will conceive of this situation as arising monomorphically, i.e. with all members of the population playing a common mixed strategy with the corresponding probabilities. The question we now ask is whether such a mixed strategy $\sigma = (p, 1-p)$ is non-invadable, i.e. whether it is an ESS (Definition 1).

Three cases may be considered.

First, the case where $V > C$. In these circumstances, to play $H$ with probability 1, i.e. to choose $\sigma = (1,0)$, is a strictly dominant strategy. Thus, no other strategy can invade it. This strategy, however, can invade any other. It defines, consequently, the only ESS in this case.

If $V = C$, to play $H$ is only weakly dominant. But, if there is any positive probability of confronting an individual playing $D$, $H$ yields a pay-off strictly higher than any other strategy, pure or mixed. Thus, again strategy $H$ is the unique ESS in this case.

Finally, consider the case where $C > V$. In these circumstances, the strategy that previously was an ESS can now be invaded by any other that gives some positive probability to playing $D$. To play $D$ with probability 1 is, nevertheless, not an ESS. For, if this strategy were adopted by the whole population, to play $H$ with some positive probability could then invade as a small-frequency mutation. Only a genuinely mixed strategy $\sigma = (p, 1-p)$ with $0 < p < 1$ can define an ESS in this case.

A first requirement that any such mixed strategy must satisfy to be a potential ESS is to yield the same pay-off to each of its constituent pure strategies. Otherwise, a lower probability to the one with a smaller pay-off could invade the population. Thus, if $p$ denotes the probability of playing $H$, such a strategy must satisfy:

$$\pi\left(H, \nu(\sigma)\right) = \pi\left(D, \nu(\sigma)\right),$$

where the population profile $\nu(\sigma) = \sigma = (p, 1-p)$ gives to each pure strategy the same weights as $\sigma$.

Since the pairing of contestants is assumed to be random, the previous equality becomes:

$$p\,\pi\left(H, (1,0)\right) + (1-p)\,\pi\left(H, (0,1)\right) = p\,\pi\left(D, (1,0)\right) + (1-p)\,\pi\left(D, (0,1)\right),$$

or
$$p\left(1/2(V - C)\right) + (1 - p)V = p \cdot 0 + (1 - p)\,V/2,$$

which yields:
$$p = V/C.$$

This value of $p$ defines a (symmetric) Nash equilibrium of the game described in Table 1: it defines a strategy that is a best response to itself. To confirm that it is indeed an ESS we need to show that no other strategy can invade it. This, according to Definition 1, requires that for any other strategy $\sigma' = (q, 1 - q)$, $q \neq p$, there should exist some $\bar{\varepsilon} > 0$ such that if $0 < \varepsilon \leq \bar{\varepsilon}$, then

$$\pi\left(\sigma, (1 - \varepsilon)\sigma + \varepsilon\sigma'\right) > \pi\left(\sigma', (1 - \varepsilon)\sigma + \varepsilon\sigma'\right).$$

which, by the linearity of $\pi$ induced by our assumption of random pairings, we can rewrite:

$$(1 - \varepsilon)\,\pi(\sigma, \sigma) + \varepsilon\,\pi(\sigma, \sigma') > (1 - \varepsilon)\,\pi(\sigma', \sigma) + \varepsilon\,\pi(\sigma', \sigma').$$

The fact that $\sigma$ is a Nash equilibrium implies that $\pi(\sigma, \sigma) \geq \pi(\sigma', \sigma)$. Thus, the above inequality is ensured if:

$$\pi(\sigma, \sigma') > \pi(\sigma', \sigma'),$$

that is, if $\sigma$ does better against $\sigma'$ than $\sigma'$ itself does. Let us confirm this inequality. It is just a matter of straightforward computation to see that:

$$\pi(\sigma, \sigma') - \pi(\sigma', \sigma') = \frac{p - q}{2}(V - qC),$$

which guarantees that, for $p \neq q$, $\pi(\sigma, \sigma') - \pi(\sigma', \sigma') > 0$. This yields the desired conclusion.

To summarize, we have been able to show that, under all values of the parameters, there exists an ESS in the Hawk–Dove game. The values of the parameters do affect, however, the nature of the equilibrium. If the resource is valuable relative to the fighting cost, the only ESS is to play Hawk with probability one. If, instead, the cost of losing a fight is higher than the value of the resource, only a completely mixed ESS exists. In it, the probability of adopting a Hawk strategy naturally depends (in a direct fashion) on the ratio of the value of the resource to the cost which may be incurred in struggling for it.

### 2.3.2  Playing the field: the Sex-Ratio game

To draw a comparison with the case of pairwise contests discussed in the previous section, we now focus on a scenario where the actual contests are population-wide (i.e. "playing the field"). The example chosen is a simplified version of the traditional sex-ratio game.

Consider some given ("infinite") population with a proportion $\zeta$ of males and $(1 - \zeta)$ of females. Each generation of females can breed a given fixed number $n$ of offspring and can mate only once. Males, on the other hand, obtain on average a number $(1 - \zeta)/\zeta$ of mates (which can obviously be greater or less than one, depending on the relative frequencies of males and females).

Not all females bear the same proportion of males in their offspring. Let us assume that only two such proportions are genetically possible: $\alpha_1 = 0.1$ and $\alpha_2 = 0.6$. They represent the two possible pure strategies (phenotypes) in the sex-ratio game we now consider.

In the simplest version of this game, females are the only relevant players and their pay-off is identified with (approximated by) the number of *grand*children they can breed. This number will depend, of course, on the sex ratio prevailing in the population during their own and their children's lifetime. Let us focus first on a situation where the whole population (except possibly the given female considered) adopts a *common* and *pure* strategy; a situation, therefore, where the population sex ratio is either $\alpha_1$ or $\alpha_2$. It can be seen immediately that, for each of the four possible instances that may a priori occur, the pay-offs for any given female adopting strategy $\alpha_i$ are as follows:

$$\pi\left(\alpha_i, (\delta_{1j}, \delta_{2j})\right) = n\left(n(1 - \alpha_i) + n\alpha_i \frac{1 - \alpha_j}{\alpha_j}\right) \qquad (2.3)$$

for each $i, j = 1, 2$, where $\delta_{kj} = 0$ if $k \neq j$, $\delta_{kj} = 1$ if $k = j$.

Normalizing, for simplicity, to $n = 1$ and particularizing for $\alpha_1 = 0.1$ and $\alpha_2 = 0.6$ we obtain the following pay-off table:

|                    |            | Population | |
| ------------------ | ---------- | ---------- | ------ |
|                    |            | $\alpha_1$ | $\alpha_2$ |
| Individual Female  | $\alpha_1$ | 1.8        | 0.967  |
|                    | $\alpha_2$ | 5.8        | 0.8    |

Table 2

From this table, it is clear that neither $\alpha_1$ nor $\alpha_2$ can, as pure strategies adopted by the whole population, become an ESS: a potential mutant does strictly better than the population by adopting the alternative strategy. Therefore, an ESS $(p^*, 1 - p^*)$, if it exists, must be mixed. That is, it must involve $p^* \in (0, 1)$.

By an argument already used in the previous section, it must then be the case that both pure strategies obtain the same pay-off when confronting a population which plays such an ESS. That is:

$$\pi\left(\alpha_1, (p^*, 1 - p^*)\right) = \pi\left(\alpha_2, (p^*, 1 - p^*)\right). \qquad (2.4)$$

Let $\alpha(p) \equiv p \cdot 0.1 + (1 - p) \cdot 0.6$ denote the average sex ratio induced by any given mixed strategy $(p, 1 - p)$ played by the whole population. Since the pay-off of any given strategy only depends on the average sex ratio it confronts, (2.4) may be rewritten as follows:

$$1 - 0.1 + 0.1\frac{1 - \alpha(p^*)}{\alpha(p^*)} = 1 - 0.6 + 0.6\frac{1 - \alpha(p^*)}{\alpha(p^*)},$$

which has $p^* = 0.2$ as its unique solution. (Note that, in contrast with the previous example with pairwise meetings, the pay-off function is no longer linear in the population profile.) The average sex ratio associated with $p^* = 0.2$ is:

$$\alpha(p^*) = 0.2 \cdot 0.1 + 0.8 \cdot 0.6 = 0.5,$$

i.e. an equal proportion of males and females.

The above condition is a necessary, but still not sufficient, condition for an ESS. To be an ESS, $(p^*, 1 - p^*)$ has to be protected from invasion, in the sense of Definition 1. This occurs if, for any other strategy which assigns a probability $q \neq p^*$ to the pure strategy $\alpha_1$, there exists some $\bar{\varepsilon} > 0$ such that if $0 < \varepsilon \leq \bar{\varepsilon}$, then

$$\pi\left((p^*, 1 - p^*), ((1 - \varepsilon)p^* + \varepsilon q, 1 - (1 - \varepsilon)p^* - \varepsilon q)\right) >$$
$$\pi\left((q, 1 - q), ((1 - \varepsilon)p^* + \varepsilon q, 1 - (1 - \varepsilon)p^* - \varepsilon q)\right). \tag{2.5}$$

Given any $\varepsilon > 0$, let $\tilde{p} \equiv (1 - \varepsilon)p^* + \varepsilon q$ be the population profile induced by an $\varepsilon$-frequency mutation towards strategy $(q, 1 - q)$ and assume, for concreteness, that $q > p^*$. (The opposite case is totally symmetric.) The corresponding average sex ratio becomes:

$$\alpha(\tilde{p}) = [(1 - \varepsilon)p^* + \varepsilon q] \cdot 0.1 + [1 - (1 - \varepsilon)p^* - \varepsilon q] \cdot 0.6,$$

which is clearly smaller than $\alpha(p^*)$. Then, from (2.3), one has:

$$\pi\left(\alpha_1, (\tilde{p}, 1 - \tilde{p})\right) < \pi\left(\alpha_2, (\tilde{p}, 1 - \tilde{p})\right).$$

Thus, since the weight that strategy $q$ assigns to $\alpha_1$ is larger than that of $p^*$, (2.5) follows, as desired.

Thus, it has been shown that in a simple sex-ratio game where females can mix in an unrestricted manner between proportions $\alpha_1 = 0.1$ and $\alpha_2 = 0.6$ of male and female offspring, a unique ESS exists. When such a strategy (by virtue of the Law of Large Numbers) is identified with the proportions actually prevailing in the large population, the average sex ratio induced by it turns out to be a one-to-one proportion of male and female offspring in each generation. It can be shown that this conclusion also holds in the general case where *all* sex proportions of offspring are, a priori, possible phenotypes.

## 2.4    ESS and Refinements of Nash Equilibrium

Consider a context with pairwise contests, as described in Subsection 2.1.2. In this section, it is first established (Proposition 1) that the ESS concept may be viewed as a refinement of Nash equilibrium in the bilateral symmetric game played by every pair of randomly matched individuals. It is further shown (Proposition 2) that it is even a refinement of one of the most common concepts used to refine Nash equilibrium: the concept of perfect equilibrium.

Denote by $A \equiv (a_{ij})_{ij=1,2,...,m}$ the $m$-dimensional pay-off matrix corresponding to the symmetric game played in every bilateral encounter.

**Proposition 1** *Let $\sigma \in \Delta^{m-1}$ be an ESS. Then,*
*(i) The pair $(\sigma,\sigma)$ is a (symmetric) Nash equilibrium of the bilateral game induced by the pay-off matrix $A$, i.e. $\forall \sigma' \in \Delta^{m-1}, \sigma \cdot A\sigma \geq \sigma' \cdot A\sigma$.*
*(ii) $\forall \sigma' \in \Delta^{m-1}, \sigma' \neq \sigma, \sigma \cdot A\sigma = \sigma' \cdot A\sigma \Rightarrow \sigma \cdot A\sigma' > \sigma' \cdot A\sigma'$.*
*Reciprocally, if (i) and (ii) hold, then $\sigma$ is an ESS.*

**Proof.** Let $\sigma$ be an ESS. If (i) fails, then:

$$\exists \sigma' \in \Delta^{m-1} : \quad \sigma' \cdot A\sigma > \sigma \cdot A\sigma,$$

which implies that, for some $\varepsilon > 0$ sufficiently small, it must be the case that:

$$\sigma' \cdot A\left((1-\varepsilon)\sigma + \varepsilon\sigma'\right) \geq \sigma \cdot A\left((1-\varepsilon)\sigma + \varepsilon\sigma'\right). \tag{2.6}$$

This contradicts Definition 1.

Suppose now that (ii) fails. That is, there is some $\sigma' \in \Delta^{m-1}, \sigma' \neq \sigma$, with

$$\sigma' \cdot A\sigma = \sigma \cdot A\sigma \tag{2.7}$$

and

$$\sigma' \cdot A\sigma' \geq \sigma \cdot A\sigma'. \tag{2.8}$$

Multiplying (2.7) by $(1-\varepsilon)$, (2.8) by $\varepsilon$, and adding both expressions we obtain (2.6), again contradicting Definition 1.

The reciprocal statement is immediate. If the inequality in (i) holds strictly, it is clear that an $\bar{\varepsilon} > 0$ can be found such that, if $\varepsilon \leq \bar{\varepsilon}$,

$$\sigma \cdot A\left((1-\varepsilon)\sigma + \varepsilon\sigma'\right) > \sigma' \cdot A\left((1-\varepsilon)\sigma + \varepsilon\sigma'\right),$$

for all $\sigma' \in \Delta^{m-1}$, $\sigma' \neq \sigma$. If (i) holds with equality for some $\sigma'$, then (ii) implies that the previous inequality must also hold in this case.    ∎

Proposition 1 establishes that the ESS concept represents a refinement of Nash equilibrium for the game played in bilateral encounters. That is, if some $\sigma$ is an ESS, it must be a Nash equilibrium (i.e. (i) above) and "something else" (satisfy condition (ii)). The next proposition indicates that the refinement of Nash equilibrium induced by the ESS concept is quite strong; in particular, stronger than that derived from the notion of perfect equilibrium.

**Proposition 2 (Bomze (1986))** *Let $\sigma \in \Delta^{m-1}$ be an ESS. Then, the pair $(\sigma, \sigma)$ is a perfect equilibrium of the bilateral game induced by the pay-off matrix $A$.*

**Proof.** It is well known (see, for example, van Damme (1987, Theorem 3.2.2)) that, in *bilateral* finite games, an equilibrium is perfect if, and only if, it involves weakly undominated strategies.

Thus, let $\sigma$ be an ESS of the bilateral game induced by the pay-off matrix $A$. Then, by Proposition 1, $(\sigma, \sigma)$ is a Nash equilibrium. Suppose, for the sake of contradiction, that $\sigma$ is a weakly dominated strategy. That is, $\exists \sigma' \in \Delta^{m-1}$ such that:

$$\forall \sigma'' \in \Delta^{m-1}, \ \sigma' \cdot A\sigma'' \geq \sigma \cdot A\sigma''.$$

Particularizing $\sigma''$ in the above expression to both $\sigma$ and $\sigma'$, we obtain the following two inequalities:

(a) $\sigma' \cdot A\sigma \geq \sigma \cdot A\sigma$;

(b) $\sigma' \cdot A\sigma' \geq \sigma \cdot A\sigma'$.

If (a) is satisfied strictly, it violates (i) in Proposition 1. Thus, suppose that it holds with equality. Then, (b) contradicts point (ii) of Proposition 1. ∎

The previous results can be strengthened even further. It can be shown, for example, that an ESS $\sigma$ defines a symmetric proper equilibrium $(\sigma, \sigma)$ of the associated bilateral game, a concept even stronger than perfect equilibrium. Additional relationships with the Nash refinement literature are thoroughly discussed in van Damme (1987).

## 2.5 The Existence of an ESS

Our primary focus in this chapter has thus far been the concept of Evolutionarily Stable Strategy. There is no available result in the literature which identifies sufficient conditions for the existence of an ESS in sufficiently general scenarios. Even when contests are assumed pairwise and random – arguably, the simplest case to analyse – straightforward examples can be provided (see below) that show that an ESS may fail to exist in quite "non-pathological" cases. However, if one wants to rule out that these examples be non-generic (i.e. dependent on precise relationships among the pay-off entries),[3] we need to consider set-ups with more than two strategies. This is established by the following generic existence result for $2 \times 2$-games:

**Proposition 3** *Consider a context of pairwise contests and two-dimensional pay-off matrix $A \equiv (a_{ij})$. Assume $a_{ii} \neq a_{ij}$ $(i = 1, 2, j \neq i)$. Then, an ESS exists.*

**Proof.** If $a_{ii} > a_{ji}$ for some $i, j = 1, 2$, $i \neq j$, the result is a consequence of the fact that in this case there exists a strict Nash equilibrium strategy which, as it is easy to see, is also an ESS. If the preceding inequality does not hold for any $i = 1, 2$, i.e. $a_{ii} < a_{ji}$ $(i = 1, 2, i \neq j)$, then the unique equilibrium (mixed) strategy of the game is $\sigma^* = (p, 1 - p)$ with $p = (a_{12} - a_{22})(a_{12} - a_{22} + a_{21} - a_{11})^{-1}$. In this case, part (ii) of Proposition 1 applies and, in order to verify that $\sigma^*$ is an ESS, it is enough to check that

$$\sigma^* \cdot A\sigma > \sigma \cdot A\sigma, \quad \forall \sigma = (q, 1 - q) \in \Delta, \ q \neq p.$$

Since

$$\sigma^* \cdot A\sigma - \sigma \cdot A\sigma = (p - q)^2 (a_{12} - a_{22} + a_{21} - a_{11}) > 0,$$

the desired conclusion follows. ∎


Despite the lack of a sufficiently general existence result, a useful sufficient condition for existence of an *interior* ESS in general contexts with pairwise encounters is provided by the following result.

**Proposition 4 (Hines (1980a))** *Let $T$ denote the tangent space of $\Delta^{m-1}$, i.e. $T \equiv \{q \in R^m : \sum_{i=1}^{m} q_i = 0\}$. Consider a context of pairwise encounters with $m$-dimensional pay-off matrix $A$ and an interior Nash equilibrium $\sigma^* \in int\left(\Delta^{m-1}\right)$. Then, if the quadratic form $q^\top \left(A + A^\top\right) q$ is negative definite for $q \in T$, $\sigma^*$ is an ESS.*

**Proof.** Since $\sigma^*$ is a Nash equilibrium and gives positive weight to all pure strategies, it follows that, for all $\sigma \in \Delta^{m-1}$ :

$$\sigma \cdot A\sigma^* = \sigma^* \cdot A\sigma^*. \tag{2.9}$$

Thus, by part (ii) of Proposition 1, $\sigma^*$ is an ESS if:

$$\sigma \cdot A\sigma < \sigma^* \cdot A\sigma \tag{2.10}$$

---

[3] This intuitive idea can be rigorously formulated: a certain set of games is said to be non-generic if, in an appropriately specified (measure) space of possible games, the closure of this set is of measure zero.

for all $\sigma \in \Delta^{m-1}$, $\sigma \neq \sigma^*$. Given any such $\sigma \in \Delta^{m-1}$, denote $x \equiv \sigma - \sigma^*$. Then, the previous expression may be rewritten as follows:

$$
\begin{aligned}
(\sigma - \sigma^* + \sigma^*) \cdot A\sigma &= x \cdot A\sigma + \sigma^* \cdot A\sigma \\
&= x \cdot Ax + x \cdot A\sigma^* + \sigma^* \cdot A\sigma \\
&< \sigma^* \cdot A\sigma.
\end{aligned}
$$

Or, using (2.9):

$$
x \cdot Ax + \sigma^* \cdot A\sigma < \sigma^* \cdot A\sigma,
$$

that is obviously equivalent to:

$$
x \cdot Ax < 0, \tag{2.11}
$$

for all $x$ of the form $\sigma - \sigma^*$ with $\sigma \in \Delta^{m-1}$, $\sigma \neq \sigma^*$. Since such $x \in T$, $x \neq 0$, the proof is complete from the assumed negative-definiteness of $A + A^\top$ in $T$. ∎

Proceeding along the lines of Proposition 4, it is a simple matter to establish the following useful *necessary* condition for the existence of an interior ESS. (See Theorem 5 for an application.)

**Proposition 5** *Consider a context of pairwise encounters with m-dimensional pay-off matrix A. If an ESS $\sigma^* \in int\left(\Delta^{m-1}\right)$ exists,[4] then the quadratic form $q^\top \left(A + A^\top\right) q$ is negative definite for $q \in T$.*

**Proof.** Let $\sigma^* \in int\left(\Delta^{m-1}\right)$ be an ESS. Then, it must satisfy (2.9) and (2.10) for every $\sigma \neq \sigma^*$. Note, moreover, that every $q \in T$ can be written as:

$$
q = \lambda\left(\sigma - \sigma^*\right),
$$

for some $\lambda \in R$ and $\sigma \in \Delta^{m-1}$. In view of (2.11) – which follows from (2.9) and (2.10) – this implies that:

$$
q \cdot Aq < 0, \quad \forall q \in T,
$$

as desired. ∎

We end this section with a simple 3-strategy scenario which starkly illustrates the problem of non-existence raised by the ESS concept. Consider a scenario with pairwise contests where the bilateral game being played is a variation of the familiar Rock-Scissors-Paper (RSP) game. There are three strategies, R, S, and P, which form a cycle in terms of the success they experience in pairwise

---

[4] If the ESS is not interior to the simplex, then the result can be reformulated with respect to the face of the simplex where it lies.

encounters: R beats S, S beats P, and P beats R. The traditional such game is zero sum: what one player wins, the other loses. This requires that if the game is to be kept symmetric, the pay-off in the case of an encounter between two identical strategies has to be made (by appropriate normalization) equal to zero.

Instead of this game we shall consider a slight variation of it in which when two identical strategies meet, they both receive a positive, albeit small, $\varphi$.[5] By equating success with a pay-off of 1, and failure with a pay-off of $-1$, the variation of the RSP game can be described by the following table:

|   | R | S | P |
|---|---|---|---|
| R | $\varphi,\varphi$ | $1,-1$ | $-1,1$ |
| S | $-1,1$ | $\varphi,\varphi$ | $1,-1$ |
| P | $1,-1$ | $-1,1$ | $\varphi,\varphi$ |

Table 3

We now argue that, in a context where every pair of randomly paired individuals plays the above bilateral game, there is no ESS. A first necessary condition for a strategy to be an ESS is that it be a best response to itself (i.e. a symmetric Nash equilibrium of the above bilateral game; cf. Proposition 1). In this game, it is immediately seen that there is only one Nash equilibrium. This is the one defined by the mixed strategy $\sigma = (1/3, 1/3, 1/3)$. Its expected pay-off is $\pi(\sigma, \sigma) = \varphi/3$.

Since $\sigma$ is a Nash equilibrium strategy, all three pure strategies in its support yield the same pay-off. Specifically:

$$\pi(R, \sigma) = \pi(S, \sigma) = \pi(P, \sigma) = \varphi/3.$$

Furthermore, the expected pay-off earned by $\sigma$ is independent of the particular (pure or mixed) strategy played by the opponent. That is:

$$\pi(\sigma, R) = \pi(\sigma, S) = \pi(\sigma, P) = \varphi/3.$$

Suppose now that some "mutants" enter the population with strategy $\sigma' = (1, 0, 0)$ in some arbitrarily small frequency $\varepsilon > 0$. Their expected pay-off is:

$$\begin{aligned} \pi(\sigma', (1-\varepsilon)\sigma + \varepsilon\sigma') &= (1-\varepsilon)\,\pi(\sigma', \sigma) + \varepsilon\,\pi(\sigma', \sigma') \\ &= (1-\varepsilon)\,\varphi/3 + \varepsilon\varphi > \varphi/3. \end{aligned}$$

Thus, strategy $\sigma'$ can invade the original $\sigma$-monomorphic population, rendering $\sigma$ evolutionary *unstable*. Since $\sigma$ is the only strategy that, being a Nash equilibrium for the bilateral game, can be a candidate for ESS, we may conclude that no ESS exists in this case.

---

[5] In fact, as explained in Subsection 3.5.2, no ESS exists even when $\varphi = 0$. Here, we postulate some $\varphi > 0$ for the sake of a more clear-cut illustration.

## 2.6 Asymmetric Contests

### 2.6.1 Introduction

Unlike what has been postulated so far, many interesting contexts of interaction in the biological world are *not* symmetric. They often involve some asymmetry among the contestants. For example, one of them may be the original owner of the resource, or the larger individual, or of different sex. Such differences may, or may not, affect pay-offs. But even if they do not, they may have strong influence on the behaviour of individuals by establishing useful correlations which may materialize as *equilibrium* conventions. An illustration of this possibility will be discussed below in the already familiar Hawk–Dove context.

The way the literature has dealt with these asymmetric situations is by embedding them into an *ex ante* symmetric context. In any particular juncture, individuals may well confront an asymmetric situation. However, their strategy cannot be restricted to just one such particular case; it has to prescribe some behaviour for each of the different cases (symmetric or asymmetric) which it may ever confront. The strategy adopted by an individual has to tell it how to behave when young or old, owner or intruder, male or female. When viewed from such a global perspective, the *contingent* strategy of an individual can be appropriately conceived as belonging to a fully symmetric game: that symmetric game in which every individual of the species faces, a priori, the same set of circumstances and with identical probability.

### 2.6.2 Ex ante symmetry with ex post asymmetries

For expositional simplicity, the presentation will focus on scenarios with pairwise random matching. In the present asymmetric context, a full description of these pairwise scenarios cannot be done just through the consideration of a single pay-off matrix $A$. It requires instead the specification of several *pairs* of matrices, one separate pair for each of the possible situations which may ever arise. We shall consider a (finite) collection of such situations, $H = R \times R$, where $R$ represents the set of possible roles to be played in each of the two "positions", 1 and 2, describing a given situation.

Consider two paired individuals, arbitrarily assigned to positions 1 and 2, with $h = (r_1, r_2) \in H$ specifying their respective roles in the given situation. Associated with situation $h$ there is a pair $(A_h, B_h)$ of pay-off matrices, $A_h$ representing the pay-off matrix of the individual in position 1 and role $r_1$, $B_h$ that of the individual in position 2 and role $r_2$. In general, of course, the situation considered need not be symmetric, i.e. the bimatrix game induced by the pair $(A_h, B_h)$ need not satisfy $A_h = (B_h)^\top$.[6] However, *ex ante* symmetry over

---

[6] Of course, both matrices $A_h$ and $B_h$ have to be of the same dimension, with the usual

*positions* does require that:

$$\forall r, r' \in R, \quad A_{(r,r')} = B^{\top}_{(r',r)}.$$

*Ex ante* symmetry also requires that the a priori possibilities for each position to occupy any of the two roles present in any given situation must be identical. To formalize this, define the density function:

$$P : H \to [0, 1],$$

reflecting the *ex ante* probability $P(h)$ of situation $h$. Symmetry over positions demands that:

$$\forall r, r' \in R, \quad P(r, r') = P(r', r). \tag{2.12}$$

Bringing together all of the above items and considerations, we define an *asymmetric context with pairwise meetings (ACPM)* as a collection

$$\Gamma = (R, P, (A_h B_h)_{h \in H}).$$

In such a description, the role of an individual is intended to summarize all the information it has in any particular situation. Therefore, any (*ex ante*) strategy in $\Gamma$ may be simply defined as a mapping

$$f : R \to A, \quad f(r) \in A(r),$$

where $A \equiv \bigcup_{r \in R} A(r)$, and $A(r)$ denotes the set of actions available to an individual in role $r$. Note that the above description allows for the possibility that there is incomplete or asymmetric information in any given situation. The exact amount of information which the individual has about the prevailing situation is fully contained in its "role". In particular, its role gives the individual enough information to know at least its set of available actions. (Of course, this set must be identical, for any given role, across all situations where this role is present. Otherwise, the implicit information this would entail would require some further differentiation of roles.)

Let $F$ denote the set of all strategies of $\Gamma$. The set of probability measures of $F$, denoted by $\Phi$, defines the set of mixed strategies of $\Gamma$. Identifying the role of an individual as its "information set", the notion of behavioural strategy can be directly transferred from classical Game Theory. Thus, associated with each mixed strategy $\phi \in \Phi$, there is a uniquely associated behavioural strategy

$$\beta : R \to \Delta(A),$$

which with every role $r \in R$ associates a probability measure $\beta(r)$ with support on $A(r)$.

As is well known, even though every mixed strategy defines a unique behavioural strategy, the converse need not be true. Such potential multiplicity

interpretation: player 1 selects rows, player 2 selects columns.

conflicts with the *strict* dominance condition required by evolutionary stability. Since, typically, there will be several mixed strategies which are "behaviourally equivalent", it will be generally impossible to single out any one of them as evolutionarily stable. Or, expressed somewhat differently, there will be no selective pressure to prevent uncontrolled drift among mixed strategies which induce the same pattern of (stochastic) behaviour for the different roles.

To remedy this problem, the notion of evolutionary stability proposed below recasts the original formulation of Definition 1 (which focused on mixed strategies) in terms of the induced behavioural strategies. Let $\beta_o$ represent the behavioural strategy monomorphically played by the population. The pay-off obtained by an ("infinitesimal") individual in position 1 which adopts some (possibly different) strategy $\beta$ may be written as follows:

$$\tilde{\pi}(\beta, \beta_o) \equiv \sum_{(r,r') \in H} P(r,r') \left[ \beta(r) \cdot A_{(r,r')} \beta_o(r') \right].$$

This leads to the following adaptation of our former concept of evolutionary stability.

**Definition 2** A *(behavioural) strategy $\beta^*$ is said to be an Evolutionarily Stable Strategy of $\Gamma$ if, given any other strategy $\beta$, there exists some $\bar{\varepsilon} > 0$ such that if $0 < \varepsilon \leq \bar{\varepsilon}$,*

$$\tilde{\pi}(\beta^*, (1-\varepsilon)\beta^* + \varepsilon\beta) > \tilde{\pi}(\beta, (1-\varepsilon)\beta^* + \varepsilon\beta).$$

Particularly interesting asymmetric contexts are those in which each one of the two individuals that meet in every encounter always play different roles (e.g. there cannot be two individuals disputing a resource who are both its former "owner", or two members of a mating couple who simultaneously play the "female" role). Such contexts are said to display *role asymmetry*.

Formally, an asymmetric context $\Gamma$ is said to display role asymmetry if

$$\forall r \in R, \quad P(r,r) = 0. \tag{2.13}$$

Besides its intuitive appeal, the preceding condition also has strong theoretical implications on the kind of strategy which may qualify as ESS. To establish these implications is the focus of the following result.

**Theorem 1 (Selten (1980))** *Let the ACPM $\Gamma$ display role asymmetry. Every ESS is a pure strategy.*

**Proof.** The key point of the proof is to show that if $\beta^*$ is an ESS, then it is also the unique best response to itself according to $\tilde{\pi}$. Thus, it has to be a pure strategy.

Suppose otherwise. Then, there exists some alternative $\beta \neq \beta^*$ such that $\tilde{\pi}(\beta, \beta^*) = \tilde{\pi}(\beta^*, \beta^*)$. Choose one of the roles in which both strategies differ, say $r^o$. Consider the strategy $\hat{\beta}$ which coincides with $\beta$ in $r^o$, and with $\beta^*$ elsewhere. Obviously

$$\tilde{\pi}(\beta^*, \beta^*) = \tilde{\pi}\left(\hat{\beta}, \beta^*\right).$$

Thus, by Proposition 1, in order for $\beta^*$ to be an ESS, it must be that:

$$\tilde{\pi}\left(\beta^*, \hat{\beta}\right) > \tilde{\pi}\left(\hat{\beta}, \hat{\beta}\right). \tag{2.14}$$

But, from the assumption of role asymmetry, $P(r^o, r^o) = 0$. Thus, when $\beta^*$ and $\hat{\beta}$ prescribe different behaviour (i.e. when the individual in question has role $r^o$), it is immaterial whether the opponent plays $\beta^*$ or $\hat{\beta}$ (since the opponent will *not* have role $r^o$, the only case in which they differ). Denote by $P(r' = r^o)$ and $P(r' \neq r^o)$ the marginal distributions induced by $P(\cdot)$ on the events $r' = r^o$ and $r' \neq r^o$, respectively, where $r'$ stands for the role of the individual in position 2. Furthermore, let $\tilde{\pi}\left(\cdot \mid r' = r^o\right)$ and $\tilde{\pi}\left(\cdot \mid r' \neq r^o\right)$ represent conditional expected pay-offs on those respective events. The previous considerations lead to the following chain of expressions:

$$
\begin{aligned}
\tilde{\pi}\left(\beta^*, \hat{\beta}\right) &= \tilde{\pi}\left(\beta^*, \hat{\beta} \mid r' \neq r^o\right) P\left(r' \neq r^o\right) \\
&\quad + \tilde{\pi}\left(\beta^*, \hat{\beta} \mid r' = r^o\right) P\left(r' = r^o\right) \\
&= \tilde{\pi}\left(\hat{\beta}, \hat{\beta} \mid r' \neq r^o\right) P\left(r' \neq r^o\right) \\
&\quad + \tilde{\pi}\left(\hat{\beta}, \hat{\beta} \mid r' = r^o\right) P\left(r' = r^o\right) \\
&= \tilde{\pi}\left(\hat{\beta}, \hat{\beta}\right),
\end{aligned}
$$

which contradicts (2.14). This completes the proof of the theorem. ∎

The previous result indicates that any ACPM which displays role asymmetry only admits pure ESS. This, of course, narrows down substantially our search for ESS, drastically simplifying the analysis of these important contexts. In general, it will also tend to aggravate the fundamental non-existence problem which already afflicts the ESS concept in symmetric environments. (Recall our discussion in Section 2.5 and see Subsection 2.6.4 below for a brief discussion on ways to circumvent this problem.)

Despite these potential problems, role asymmetry in an ACPM may lead to new and very interesting features which are only possible when the underlying context is sufficiently rich to allow for *ex post asymmetries*. An illustration to this effect is included in the next subsection.

### 2.6.3 Example: the Hawk–Dove game revisited (I)

In Subsection 2.3.1, we discussed the popular Hawk–Dove game in its original (symmetric) formulation. Presently, we turn to an asymmetric version of it, whose conclusions are qualitatively very different.

Consider a Hawk–Dove context as described above, except that we now incorporate the following *pay-off-irrelevant* feature. When two individuals meet they occupy different roles: owner ($o$) and intruder ($i$). Neither of these roles affects the pay-offs of the interaction which are as in Table 1.

This defines an asymmetric game

$$\Gamma = \left(R, P, (A_h, B_h)_{h \in H}\right),$$

where:

$$R = \{o, i\}$$

and

$$\forall h \in H \equiv R \times R, \ A_h = (B_h)^\top.$$

Furthermore, by (2.12) and role asymmetry,

$$P(o, i) = P(i, o) = 1/2.$$

From Theorem 1, we know that only a pure (behavioural) strategy of the type:

$$\beta : \{o, i\} \to \{H, D\},$$

can be an ESS.

If $V > C$, it is clear that, as in the symmetric context, the constant strategy $\beta(r) = H$ is the only ESS.[7]

If $V < C$, the strategy

$$\beta^*(o) = H; \quad \beta^*(i) = D, \tag{2.15}$$

is easily verified to be an ESS. In this case, the owner exploits the asymmetry of roles in the game to obtain the *full* value of the resource. But, equivalently, roles could be used to co-ordinate actions in the opposite way. That is, the strategy

$$\hat{\beta}(o) = D; \quad \hat{\beta}(i) = H, \tag{2.16}$$

is also an ESS of the game. Since the *pay-off* structure of the game is fully symmetric across roles, these can be used in totally equivalent ways to co-ordinate actions.

In a sense, the situation is very reminiscent of the concept of *correlated equilibrium* studied by classical Game Theory. Here, it is the *ex ante* lottery over roles which serves to achieve the *equilibrium* correlation between players'

---

[7] When $V = C$, there is no ESS.

actions. And indeed, as could also happen with a suitably selected correlated equilibrium, the present correlation improves the expected (or average) pay-offs of the individuals involved.

To see this, note that in any of the two ESS's defined by (2.15) or (2.16), individuals obtain an expected pay-off equal to:

$$\frac{1}{2} \cdot V + \frac{1}{2} \cdot 0 = \frac{V}{2}.$$

In contrast, the pay-off obtained in the symmetric Hawk–Dove game studied in Subsection 2.3.1 is:

$$\frac{V}{C} \cdot \frac{V-C}{2} + \left(1 - \frac{V}{C}\right) \cdot V = \frac{V}{2} \left(1 - \frac{V}{C}\right).$$

Clearly, the latter pay-off is always smaller than the former if $V < C$.

### 2.6.4    *Extensive-form contests*

Games represented in extensive form are always formally asymmetric since the extensive form *representation* of a game always involves an unavoidable asymmetry. Even if it formalizes a *simultaneous* interaction, there is always a well-defined (but fictitious) order of move, players moving at subsequent stages of the game assumed uninformed of the choices made by their predecessors.

Many such asymmetries can be seen as purely a matter of representation and, therefore, circumvented by appropriate identification of equivalent representations (see Selten 1983, 1988). However, when this is the case and the game may be viewed as essentially symmetric, the extensive-form representation is an unnecessarily complex object to represent it. The alternative representation in normal (or strategic) form does not lose any essential information and is a much more amenable object of analysis. It may then be analysed as explained in the preceding sections.

In contexts where the game in extensive form genuinely describes a multi-stage situation (i.e. it involves a real ordering of moves), the analysis of the evolutionary stability of its equilibria brings up issues quite reminiscent of those studied above for general asymmetric conflicts. They will merely be sketched here.

On the one hand, there is the issue that, as in asymmetric games, evolutionarily stable strategies in extensive-form games will tend to be pure, i.e. will prescribe deterministic choices at each information set. As shown by Selten (1983), this will occur for all those information sets which, heuristically, already induce some asymmetry in the future "roles" of the player for the remaining part of the game. (As explained, all genuinely dynamic games will *eventually* display such asymmetry.) The choice of the term "role" in this statement is already indicative of the rationale, very analogous to that used in Theorem 1 above, which leads to the previous conclusion.

On the other hand, there is a problem similar to the lack of selective pressure on equivalent behaviour discussed above for asymmetric conflicts. In the present context, the problem manifests itself in the fact that, in extensive-form games, some subgames may never be reached in equilibrium. Thus, what the strategy prescribes in those subgames can have no evolutionary relevance. In particular, evolutionary forces will be incapable of preventing uncontrolled drift among such equivalent strategies, even if they are modelled in a behavioural form. None of them can, therefore, be evolutionarily stable in a strict sense.

To remedy this problem, Selten (1983, 1988) proposed the concept of *limit ESS*.[8] Essentially, it reflects the idea that the equilibrium strategy may be seen as the limit of evolutionarily stable strategies for close "perturbed games". In these perturbed games, admissible strategies may be required to play some actions with arbitrarily small (but positive) probability, thus ensuring that subgames which might otherwise not be reached in the original game are visited with at least some positive probability. In this fashion, one overcomes the key cause for non-existence outlined above, i.e. the fact that evolutionary forces prove ineffective at unreached subgames.

In a sense, the approach underlying the idea of a limit ESS is quite reminiscent of that associated with perfect equilibrium, a concept also proposed by Selten himself. However, it is important to understand that, unlike this latter concept (whose motivation is to *refine* the concept of Nash equilibrium), the rationale for the present one is to *generalize* the ESS concept in order to obtain the desired existence. Thus, in particular, every ESS is always itself a limit ESS since one may always consider the original game as a close and "degenerate" perturbed game. This contrasts with the approach underlying the concept of perfect equilibrium where such degenerate perturbations are not allowed: *all* perturbed games must have *every* pure strategy played with some positive probability.

## 2.7 ESS and Finite Populations

### 2.7.1 The "spite" of an ESS

The theoretical framework considered so far has, more or less implicitly, focused on a context where there is an "infinite" population, i.e. one whose members are infinitesimal relative to the size of the population.

This was the case, for example, in our discussion of the ESS concept in Section 2.2. Recall that, in interpreting this definition, a crucial point was that, in the event of a mutation with $\varepsilon$ frequency, both the predominant phenotype and the mutant confront the same population profile. But, if the population

---

[8] See Samuelson (1990) for an application of this concept to asymmetric contexts. As explained above, these contexts are subject to similar non-existence problems.

is finite, this is clearly incorrect. (Heuristically, the "$\varepsilon$-frequency" of mutants which the majority of the population faces is not the same, i.e. is greater, than that of the mutants themselves.) And, as we shall see, the issue is not just one of formal inaccuracy: if the population is relatively small, it may have significant implications as to what has to be viewed as evolutionarily stable.

Consider a context as described in Section 2.1, except that now the (finite) population is composed of $n$ individuals. Let $\sigma^*$ denote the strategy adopted by some originally monomorphic population, and let $\rho$ be the strategy of a mutant threatening to invade the population. The intuitive idea underlying the issue of evolutionary stability is as before: Can $\sigma^*$ resist and expel the mutant $\rho$? The following definition captures the corresponding notion of non-invadability, as applied to the present finite-population context.[9]

**Definition 3** *A strategy $\sigma^* \in \Delta^{m-1}$ is said to be evolutionarily stable if, for any other strategy $\rho$ and $\varepsilon = 1/(n-1)$,*

$$\pi\left(\sigma^*, (1-\varepsilon)\sigma^* + \varepsilon\rho\right) > \pi\left(\rho, \sigma^*\right).$$

The previous definition evaluates a situation where an originally monomorphic population playing $\sigma^*$ is threatened by a single mutation $\rho$. In order to have $\sigma^*$ evolutionarily stable, it is necessary that its associated expected pay-off (when confronting a profile with $n-2$ non-mutants and one single mutant)[10] is higher than that corresponding to strategy $\rho$ when facing a *homogeneous* profile of $n-1$ non-mutants. That is, only non-mutants confront the (single) mutant with probability $1/(n\text{-}1)$; the latter only confronts non-mutants.

There are two implicit but crucial assumptions underlying Definition 3.

One is that mutations are so rare that they appear only one at a time (and in just a single individual). A similar idea lay behind the infinite-population version of ESS presented in Definition 1. As explained in the following chapters, our conclusions (those here as well as the preceding ones) may be substantially affected if several mutations may arise at the same time.

A second implicit assumption is that the appropriate pay-offs in order to evaluate the survival possibilities of a given strategy are expected pay-offs. When the population is very large (i.e. formally infinite), such an assumption is a good enough approximation since expected and average pay-offs will tend to come very close. However, if the population is relatively small (formally finite), there will typically be large amounts of noise present in the interaction. This brings important issues into the picture, whose analysis is again postponed to ensuing chapters. For the moment, the identification of average and expected pay-offs will be maintained, even for finite populations.

To illustrate the implications of the ESS concept in finite populations, let us restrict our attention to contexts with pairwise random meetings and a bilateral

---

[9] Much of what is discussed in this subsection is taken from Shaffer (1988, 1989).

[10] As a natural adaptation of the infinite-population concept, the population profile confronted by a given individual is identified with the strategy frequencies played by the rest of the population.

symmetric game with pay-off matrix $A$. The inequality in Definition 3 may then be formulated as follows:

$$\tfrac{n-2}{n-1}\pi\left(\sigma^*,\sigma^*\right) + \tfrac{1}{n-1}\pi\left(\sigma^*,\rho\right) > \pi\left(\rho,\sigma^*\right). \tag{2.17}$$

For arbitrarily large $n$, a strategy $\sigma^*$ satisfies (2.17) only if it is a Nash equilibrium of the bilateral game. This is in accord with Proposition 1, which referred to contexts with an infinite population.

For $n$ relatively small, however, a strategy $\sigma^*$ satisfying (2.17) for all $\rho \neq \sigma^*$ may be far from being a symmetric Nash equilibrium of the bilateral game, i.e. it may not be a best response to itself. This can best be seen if (2.17) is rewritten as

$$\begin{aligned} \pi\left(\sigma^*,\sigma^*\right) \ & > \tfrac{n-1}{n-2}\pi\left(\rho,\sigma^*\right) - \tfrac{1}{n-2}\pi\left(\sigma^*,\rho\right) \\ & = \pi\left(\rho,\sigma^*\right) - \tfrac{1}{n-2}\left[\pi\left(\sigma^*,\rho\right) - \pi\left(\rho,\sigma^*\right)\right]. \end{aligned} \tag{2.18}$$

This expression indicates that, for strategy $\sigma^*$ to be able to expel any given mutant $\rho$, the former need not be an optimal response to itself. In other words, even if some mutant $\rho$ is a *strict* better response to $\sigma^*$, i.e. $\pi\left(\rho,\sigma^*\right) > \pi\left(\sigma^*,\sigma^*\right)$, (2.17) (or (2.18)) may still hold if $\pi\left(\sigma^*,\rho\right) - \pi\left(\rho,\sigma^*\right)$ is sufficiently large relative to the size of the population.

The previous expression reflects the fact that, in finite populations, a successful mutation not only has to compare its pay-off with the original one of the predominant strategy, i.e. compare $\pi\left(\rho,\sigma^*\right)$ and $\pi\left(\sigma^*,\sigma^*\right)$. In addition, it must take into account the effect that its entrance will have on the pay-off of the prevailing strategy as compared to its own pay-off. When this difference is appropriately scaled by the size of the population, we are led to the term $\left(\pi\left(\sigma^*,\rho\right) - \pi\left(\rho,\sigma^*\right)\right)/\left(n-2\right)$ appearing in (2.18).

This may be better understood by adopting a reciprocal perspective and asking what makes a population-wide strategy resist any given mutation threat. As indicated by the above expression, a certain strategy *not* defining a symmetric Nash equilibrium may still be evolutionarily stable if any deviation from it which is profitable (à la Nash) would nevertheless benefit the dominant phenotype in *relative* terms, i.e. even more than the mutant. This "concern" for *relative* (rather than absolute) pay-offs is what, in another context, Hamilton (1970) has labelled *spiteful behaviour*. This type of behaviour represents a distinctive feature of the ESS concept in finite populations.

### 2.7.2   *An example of oligopolistic competition*

Consider a given industry composed of $n$ firms, all of which sell a certain homogeneous product.[11] Let the demand function of this product be given

---

[11] This example was developed jointly with Luis Corchón.

by a certain (strictly) decreasing function $P(\cdot)$, whose argument is the sum $x_1 + x_2 + ... + x_n$, where each $x_i$ denotes the sales of firm $i = 1, 2, ..., n$. Let cost conditions be identical for each firm $i$ and represented by a common (differentiable) cost function $C(\cdot)$ with $\frac{d^2C(\cdot)}{d(x_i)^2} > 0$.

Suppose firms choose their quantities simultaneously. A *symmetric* Cournot-Nash equilibrium of this game, $(x^*, x^*, ..., x^*)$ is characterized by the inequality:

$$P(n\,x^*)\ x^* - C(x^*) \geq P((n-1)\,x^* + x)\ x - C(x),$$

for all $x \geq 0$.

On the other hand, Walrasian (or competitive) behaviour requires that each firm $i = 1, 2, ..., n$ chooses a quantity $x^w$ such that:

$$P(n\,x^w)\ x^w - C(x^w) \geq P(n\,x^w)\ x - C(x), \tag{2.19}$$

for all $x \geq 0$. That is, each firm takes the market-clearing price $P(n\,x^w)$ as given and maximizes profits. Under our assumptions on $P(\cdot)$ and $C(\cdot)$, it is easy to see that $x^* < x^w$. Thus, in particular, $(x^w, x^w, ..., x^w)$ is *not* a Nash equilibrium.

It is now shown that $x^w$ is, however, an ESS for the $n$-firm industry described. Specifically, if all other firms choose $x^w$, any single firm which chooses a different output always obtains a strictly lower profit than the other $n-1$ firms. (This profit, of course, can be greater *in absolute terms* than what it would have obtained by choosing $x^w$, given that $(x^w, ..., x^w)$ is not a Nash equilibrium.) Thus, it is claimed that the following inequality holds:

$$P((n-1)\ x^w + x)\ x^w - C(x^w) > P((n-1)\ x^w + x)\ x - C(x), \tag{2.20}$$

for all $x \neq x^w$.

Since $P(\cdot)$ is decreasing, we have:

$$[P(n\,x^w) - P((n-1)\ x^w + x)]\ x^w < [P(n\,x^w) - P((n-1)\ x^w + x)]\ x,$$

or

$$P(n\,x^w)(x^w - x) < P((n-1)\ x^w + x)(x^w - x).$$

Subtracting the term $[C(x) + C(x^w)]$ from both sides, and after some rearrangement, we get:

$$[P((n-1)x^w + x)\ x^w - C(x^w)] + [P(n\,x^w)\ x - C(x)] >$$
$$[P((n-1)\ x^w + x)\ x - C(x)] + [P(n\,x^w)\ x^w - C(x^w)].$$

By (2.19) the second term of the right-hand side (RHS) of the above expression is no smaller than the corresponding second term of the left-hand side (LHS). Thus, it follows that the first term of the LHS is larger than the corresponding first term of the RHS. This yields expression (2.20), as desired.

The previous argument shows that the Walrasian strategy (output) $x^w$, as defined by (2.19), is evolutionarily stable. Thus, if survival is linked to differential profits, a firm which lives in a competitive industry will only survive by being competitive.

It is important to emphasize that the previous conclusion is generally false if other firms are not competitive. Suppose, however, that either $n = 2$ or that, for arbitrary $n$, the context is one of random bilateral encounters. Then, the above argument indicates that a firm will survive (i.e. will obtain at least as large profits as its competitors) by choosing the Walrasian output, *irrespective* of the strategies chosen by the other firms. In a certain sense, we may say that, in this context, Walrasian behaviour is a "dominant strategy" for survival.

## 2.8 Evolution and Cheap Talk

We end this chapter with an interesting application of evolutionary analysis. It involves a natural rationalization of "cheap talk" as a mechanism for the consolidation of co-operative (i.e. efficient) behaviour.[12]

Besides its inherent interest, this application is also quite useful for its illustrative potential. In a single set-up, it elaborates many of the different issues studied throughout this chapter. For example, it will bear on the topic of monomorphic vs. polymorphic interpretations of an ESS which was discussed in Subsection 2.2.2. It will also focus on an asymmetric and multi-stage context, as considered in Section 2.6. Finally, it will involve a finite population of individuals, as studied in Section 2.7.

The model proposed here is largely inspired by Sobel (1993). (See also Matsui (1991), Wärneryd (1993), Bhaskar (1994), or Kim and Sobel (1995) for a discussion of related issues.) Consider a finite population of $n$ agents playing a certain bimatrix game $G$ with pay-off matrices $A$ and $B$. Individuals are randomly paired to play the game, occupying one of the two alternative positions in it, 1 or 2, with the same probability. If an individual occupies position 1, he is identified with the "row player", his pay-off matrix is $A$, and his action set is $Q_1$. If he occupies position 2, his pay-off matrix is $B$, and his strategy set is $Q_2$. For notational simplicity, we find it convenient to extend the previous notation and write $\tilde{\pi}_k(a_1, a_2)$ as the pay-off achieved by a player in position $k \in \{1, 2\}$ when the individual in position 1 plays action $a_1$ and that in position 2 plays action $a_2$.[13]

Assuming that the number of matchings taking place is very large, we shall identify the *total* realized pay-off of playing any strategy against a certain pop-

---

[12] I am indebted to Joel Sobel for very helpful conversations in the preparation of this section.

[13] Note that, in terms of the context for an ACPM developed in Section 2.6, there is no role diversity in each position (i.e. we can think of just having a single role in each position, the game asymmetry being exclusively linked to the existence of alternative player positions).

ulation profile with its *expected* pay-off. (Equivalently, we may think of the situation as being one where players are involved in a round-robin tournament, each of them playing everyone else once in each of the two roles.) In this context, the full set of possible strategies is the collection of all those contingent plans which specify an action to be played in each of the two roles. Thus, formally, it can be identified with the set $S \equiv Q_1 \times Q_2$, whose typical element $s = (a_1, a_2)$ indicates the actions $a_1$ and $a_2$ to be played in the first and second roles, respectively.

The particular kind of game we shall consider here is usually referred to as a *game of common interest*. As its label indicates, this is a game where the players are subject to no conflict of interest whatsoever: the outcome one of them prefers in any given role is also the outcome the other prefers in the alternative role. Formally, this idea is contained in the following definition.[14]

**Definition 4** *A bimatrix game $G$ as described is called of common interest if there exists a single action pair $(a_1^*, a_2^*) \in Q_1 \times Q_2$ such that $\tilde{\pi}_k(a_1^*, a_2^*) \geq \tilde{\pi}_k(a_1, a_2), \forall k = 1, 2, \forall (a_1, a_2) \in Q_1 \times Q_2$.*

For the sake of illustration, suppose that the game $G$ is a simple pure co-ordination game with just two actions and two corresponding pure-strategy equilibria. In the first equilibrium, $(\hat{a}_1, \hat{a}_2)$, the induced pair of pay-offs is $(\hat{\pi}_1, \hat{\pi}_2)$; in the second equilibrium, $(\tilde{a}_1, \tilde{a}_2)$, the induced pair of pay-offs is $(\tilde{\pi}_1, \tilde{\pi}_2)$. Further assume that, for each $k = 1, 2, \hat{\pi}_k > \tilde{\pi}_k$, i.e. the first equilibrium dominates the second one in the Pareto sense. Despite this unambiguous ranking of the two equilibria, it is clear that the following two strategies of the asymmetric context, $\hat{s} = (\hat{a}_1, \hat{a}_2)$ and $\tilde{s} = (\tilde{a}_1, \tilde{a}_2)$, *both* define an *evolutionarily stable strategy* in the sense of Definition 3. This illustrates the fact that, unless one enriches the underlying framework appropriately, there is no hope of finding a purely evolutionary argument which, even in simple games of common interest, would select for efficient configurations.

A natural way of enriching the context in cases where no conflict of interest exists is to allow for communication among the players. In those contexts where all players share the same objectives, it would seem that unrestricted communication among them should naturally lead to co-ordination on their commonly most preferred outcome (i.e. strategy profile).

In addressing this issue, the first choice to make is a modelling one. How are we to model such imprecise notions as "communication"? Even though much more complicated formalizations of the idea would perform similarly, the literature has often relied on the following simple one. Every pair of players who are matched becomes involved in a two-stage game. In the first stage, they simultaneously decide on a message $m \in M$ to send to their opponent. (We shall assume that the message space is sufficiently rich so that $|M| > n$, i.e.

---

[14] This is stronger than the concept proposed by Aumann and Sorin (1989) in that these authors admit that different action profiles may lead to the same most preferred pay-off profile. This slightly more general concept could have been adopted at the cost of some complication in the argument.

there are more messages than individuals.) The messages are costless, in the sense that they will not play any direct role (either positive or negative) in the eventual pay-offs earned by the players. The only *indirect* way in which they may impinge on players' pay-offs is by affecting what they decide to do in the second stage of the game. In this second stage, players are confronted with a game $G$ as described above. In deciding upon the continuation strategy to play at this point, players are assumed to have observed the message sent by their opponent (and remember also its own message), possibly linking their choice at this stage to these observed messages.

In the two-stage game with pre-play communication described, strategies are somewhat complicated. On the one hand, they must specify the message $m_k \in M$ to be sent in the first stage of the game, depending on the position $k = 1, 2$ that the individual occupies in the game.[15] In addition, they must determine what action to choose, given the pair of messages $(m_1, m_2)$ observed, again depending on the position $k$ occupied. Since there are $|M|^2$ such possible pairs of message observations, a complete strategy for the communication game can be identified with an element of the set $\Phi \equiv M^2 \times (S^{|M| \times |M|})^2$.

It is easy to see that it is impossible in this context to find a single pure strategy profile that could alone qualify as evolutionarily stable. The reason is akin to the ideas verbally explained in Subsection 2.6.4. If a single pure strategy profile is played, all but two of the action-message components contained in each player's strategy are irrelevant. Therefore, there is no selective pressure on them which may render any *particular* choice stable.

These considerations suggest extending our former notions of evolutionary stability to a set-valued concept which allows for the co-existence of a variety of alternative strategies played by different individuals. This is, for example, the approach undertaken by Sobel (1993), which captures the idea that agents may "drift" by adopting (in a weakly optimal fashion) new strategies which only differ from the original ones in contingencies (i.e. messages) which are irrelevant (i.e. currently not sent). Here, we focus on a very similar approach. However, unlike Sobel's, it is evolutionary in a strict "biological sense". That is, as in the concepts discussed so far (e.g. ESS), it evaluates the viability of any new strategies on the exclusive basis of their *relative* pay-offs. (It ignores, therefore, any considerations related to agents' best responses *per se*.) To proceed formally, some additional notation is first introduced.

Let $\Phi^n$ be the set of strategy profiles in the communication game. A typical element of it will be denoted by $\underline{\phi} \equiv (\phi^1, \phi^2, ..., \phi^n)$. For each $i = 1, 2, ..., n$, $\phi^i = [(m_1^i, m_2^i), (\varphi_1^i(\cdot), \varphi_2^i(\cdot))] \in \Phi$ denotes the strategy played by individual $i$, its interpretation being as explained above. We shall use the notation $\underline{\phi} \setminus_i \hat{\phi}$ to represent the profile of strategies resulting when the original $\phi^i$ played by

---

[15] We are implicitly assuming that the message exchange takes place after each player learns his position in the game. Alternatively, it could have been assumed that the position assignment is only resolved after the message exchange has taken place. Nothing essential would be affected by this alternative formulation.

individual $i$ is replaced by $\hat{\phi}$. For any given $\underline{\phi}$, the pay-off obtained by player $i$ is given by some function $\pi^i(\underline{\phi})$ which reflects the expected (or average) magnitudes resulting from the postulated matching framework. Specifically, it is defined as follows:

$$\pi^i(\underline{\phi}) = \sum_{j \neq i} \tilde{\pi}_1(\varphi_1^i(m_1^i, m_2^j), \varphi_2^j(m_1^i, m_2^j)) + \sum_{j \neq i} \tilde{\pi}_2(\varphi_1^j(m_1^j, m_2^i), \varphi_2^i(m_1^j, m_2^i)).$$

$$(2.21)$$

The previous expression simply reflects the assumption that every individual $i = 1, 2, ..., n$ meets (or is expected to meet) every other individual $j \neq i$ exactly once (respectively, with uniform probability) in each of the two game positions.[16]

The evolutionary concept proposed is described in the following definition.[17]

**Definition 5** *A set $\Theta \subset \Phi^n$ is an evolutionarily absorbing set (EAS) if it is a non-empty set of strategy profiles which is minimal with respect to the following twofold property:*

$$\forall \underline{\phi} \equiv (\phi^1, \phi^2, ..., \phi^n) \in \Theta, \ \forall i = 1, ..., n,$$

*(a)* $\left[ \exists j \in \{1, ..., n\} : \ \pi^i(\underline{\phi}) < \pi^j(\underline{\phi}) \right] \Rightarrow \underline{\phi} \setminus_i \phi^j \in \Theta.$

*(b)* $\left[ \exists \hat{\phi} \in \Phi \ s.t. \ \forall j = 1, ..., n, \ \pi^i(\underline{\phi} \setminus_i \hat{\phi}) \geq \pi^j(\underline{\phi} \setminus_i \hat{\phi}) \right] \Rightarrow \underline{\phi} \setminus_i \hat{\phi} \in \Theta.$

$$(2.22)$$

The twofold property contemplated by the previous definition implicitly embodies the two evolutionary forces which interest us here. Part (a) captures selection forces: if the strategy $\phi^i$ adopted by individual $i$ earns a pay-off lower than that of some opponent, then evolution may replace $i$'s strategy with that of the opponent. Part (b), on the other hand, reflects the dynamics of drift: if any "mutation" $\hat{\phi}$ renders player $i$ a pay-off as great as that of any other agent, then it may evolve into the set $\Theta$.

The previous concept could be modified in a number of ways without affecting the analysis. For example, part (b) could be reformulated so that a strategy which earns at least average pay-offs may evolve into the set $\Theta$. Analogously, part (a) could be modified so that if a strategy fares worse than average (i.e. not just worse than *some* strategy) it is eligible for replacement by the strategy of a better-off opponent. Of all the different possible variations which could be

---

[16] Strictly speaking, the formulation contained in (2.21) describes the round-robin interpretation of our matching framework. If the random-matching alternative interpretation were considered, the above expression would have to be multiplied by the constant $1/[2(n-1)]$.

[17] Besides Sobel's (1993) similar concept mentioned above, other related notions have been proposed in the literature. For example, Thomas (1985) proposes an alternative set-valued concept which requires that any invading strategy renders a strictly better pay-off than the one it replaces. Swinkels (1992) strengthens the previous requirement by postulating that entrants play an optimal strategy relative to the post-entry configuration. Both of them focus on sets of Nash equilibria which are closed with respect to their respective entry condition.

contemplated, the arguably simplest one has been chosen in Definition 5 for the sake of clarity.

The considerations underlying a EAS are qualitatively different from the *strict* selection pressures which sustain a given ESS. As will be recalled, an ESS is required to dominate strictly any other alternative strategy which might enter the population in small frequency. The EAS concept is more "pragmatic": if an alternative strategy to be played by some member of the population yields a pay-off as large as that of all others, it is admitted. Note, however, that EAS generalizes the ESS concept, as defined in Definition 3 for finite populations. That is, any ESS (when one exists) obviously defines a *singleton* EAS.

When an EAS is not a singleton (as will be the case here), this concept provides a theoretical underpinning for the sort of polymorphism which (as explained in Subsection 2.2.2) is inconsistent with the implicit motivation underlying the ESS concept. However, along the lines of some of our former discussion, one could still argue that the static nature of this new concept should provide little insight into phenomena such as polymorphism and drift, which are inherently dynamic. Even though this might be a reasonable criticism in general, it loses much of its appeal in those particular cases where every element of a certain "evolutionarily stable set" induces the same kind of realized behaviour. This is indeed the state of affairs established by the central result of this section.

**Theorem 2** *Consider a common interest game G subject to pre-play communication as described. A unique EAS $\Theta$ exists. Moreover, if the population is large enough, $\forall(\phi^1, \phi^2, ..., \phi^n) \in \Theta, \forall i, j = 1, ..., n \ (i \neq j), \ \varphi_1^i(m_1^i, m_2^j) = a_1^*$ and $\varphi_2^i(m_1^j, m_2^i) = a_2^*$.*

**Proof.** First, it is argued that an EAS always exists. Since this concept is defined as the minimal non-empty set satisfying condition (2.22), it is enough to show that a non-empty set satisfying this condition exists. But, obviously, the whole set $\Phi^n$ satisfies it, thus confirming existence of an EAS.

Next, it is claimed that $\Theta$ must have all its strategy profiles $\underline{\phi} \in \Theta$ satisfy:

$$\forall \underline{\phi} \in \Theta, \forall i, j = 1, ..., n \ (i \neq j), \ \varphi_1^i(m_1^i, m_2^j) = a_1^*, \ \varphi_2^i(m_1^j, m_2^i) = a_2^*,$$
(2.23)

i.e. they have to induce the (single) efficient pair of actions. To show this, we argue that any profile of strategies which does *not* have this property cannot be part of an EAS. Let $^0\underline{\phi}$ be some such strategy profile. The argument involves two parts.

First, it is shown that a finite chain of strategy profiles $(^0\underline{\phi}, ^1\underline{\phi}, ..., ^r\underline{\phi})$ can be constructed where $^r\underline{\phi}$ satisfies condition (2.23) and $\forall q = 1, ..., r, \ \exists i$ such that:

$$\forall k \neq i. \ ^{(q-1)}\phi^k = \ ^q\phi^k.$$
(2.24)

and either[18]

$$\pi^i({}^{(q-1)}\underline{\phi}\setminus_i({}^q\phi^i)) \geq \pi^j({}^{(q-1)}\underline{\phi}\setminus_i({}^q\phi^i)), \quad \forall j = 1, 2, ..., n, \qquad (2.25)$$

or

$$q\phi^i = {}^{(q-1)}\phi^j, \qquad (2.26)$$

with

$$\pi^i({}^{(q-1)}\underline{\phi}) < \pi^j({}^{(q-1)}\underline{\phi}). \qquad (2.27)$$

Secondly, it will be shown that no converse transition is possible, i.e. there is no path satisfying (2.24)–(2.27) which, from a strategy profile that satisfies (2.23), may lead to another strategy profile *not* satisfying it.

To show the first part, let $\hat{m} \in M$ be some message not included in the strategy profile ${}^0\phi$ by any player in one of the positions, say 1. Some such message is bound to exist since a rich enough message space has been assumed ($|M| > n$). It is now claimed that, from ${}^0\phi$, one can construct a chain satisfying (2.24)–(2.27) which after $n$ steps at most has all players in position 2 reacting to a (hypothetical) message $\hat{m}$ by its opponent with action $a_2^*$.

The intuitive reason is that, since $\hat{m}$ is never sent by players in position 1 along the contemplated chain, any reaction to it has no pay-off relevance. Formally, for each $q = 1, 2, ..., n$, denote:

$$R(q) = \{i : \pi^i({}^q\underline{\phi}) = \max_{k=1,...,n} \pi^k({}^q\underline{\phi})\}$$

$$U(q) = \{i : {}^q\varphi_2^i(\hat{m}, \cdot) = a_2^*\}.$$

Then, at each $q = 1, 2, ..., r_1$ ($r_1 \leq n$), consider the following transitions:

(i)  If $R(q-1)\setminus U(q-1) \neq \emptyset$, choose any $i \in R(q-1)\setminus U(q-1)$ and change its strategy at $q$ so that ${}^q\varphi_2^i(\hat{m}, \cdot) = a_2^*$.

(ii) Otherwise (i.e. if $R(q-1)\setminus U(q-1) = \emptyset$), choose any $i \notin U(q-1)$ and make ${}^q\phi^i$ equal to the strategy ${}^{(q-1)}\phi^j$ of some individual $j \in R(q-1)$.

Note that if (i) applies, it induces an admissible transition due to (2.24) and (2.25). If, on the other hand, (ii) applies, it also leads to an admissible transition since, in this case (i.e. provided (i) does *not* apply), the contemplated agents $i$ and $j$ must satisfy $i \notin R(q-1)$ and $j \in U(q-1)$. Thus, the postulated transition is consistent with (2.24) and (2.26)–(2.27). Clearly, in $n$ steps at most (i.e. with $r_1 \leq n$), we have that

$$r_1\varphi_2^i(\hat{m}, \cdot) = a_2^*, \, \forall i = 1, 2, ..., n. \qquad (2.28)$$

Consider now some message $\tilde{m} \in M$ not used by any player in position 2 when profile ${}^{r_1}\underline{\phi}$ prevails, and consider an analogous consecutive chain of

---

[18] Note, of course, that ${}^{(q-1)}\phi\setminus_i({}^q\phi^i) = {}^q\underline{\phi}$.

profiles for $q = r_1 + 1, ..., r_2$. This chain may be constructed so that condition (2.28) is not altered throughout, but at $q = r_2$ one has:

$$^{r_2}\varphi_1^i(\cdot, \tilde{m}) = a_1^*, \ \forall i = 1, 2, ..., n, \tag{2.29}$$

i.e. the analogue of (2.28) for message $\tilde{m}$. Now simply consider a final chain for $q = r_2 + 1, ..., r_2 + n \ (\equiv r)$ which for each $i = q - r_2$ has individual $i$'s strategy (and only this strategy) changed to satisfy

$$^q m_1^i = \hat{m}; \quad {}^q m_2^i = \tilde{m}$$

and

$$^q\varphi_1^i(\hat{m}, \cdot) = a_1^*; \quad {}^q\varphi_2^i(\cdot, \tilde{m}) = a_2^*,$$

without violating (2.28) and (2.29). Such transitions are admissible by virtue of (2.25) and the eventual profile $^r\phi$ clearly satisfies (2.23). This completes the first part of the proof.

Consider now any strategy profile $\phi$ which satisfies (2.23). Since all individuals are attaining at $\phi$ the same pay-off, it just needs to be checked that no transition of the type contemplated in (b) of Definition 5 may lead the population to violate (2.23). Any *unilateral* change by some individual $i$ to a strategy $\hat{\phi}^i \neq \phi^i$ which leads to such a violation must decrease $i$'s pay-off by some minimum amount $\Delta$ (which depends on the pay-offs of the game) for *each* one of its encounters in which an action profile different from $(a_1^*, a_2^*)$ occurs. However, only if there are at least $n - 1$ such encounters could *every* individual $j \neq i$ obtain a pay-off lower than the maximum pay-off

$$\pi^* \equiv (n - 1) \left[ \pi_1(a_1^*, a_2^*) + \pi_2(a_1^*, a_2^*) \right].$$

And then, the difference with this pay-off could be no larger than

$$\begin{aligned} \eta \ \equiv \ & \left[ \pi_1(a_1^*, a_2^*) - \min_{(a_1, a_2) \in Q_1 \times Q_2} \pi_1(a_1, a_2) \right] \\ & + \left[ \pi_2(a_1^*, a_2^*) - \min_{(a_1, a_2) \in Q_1 \times Q_2} \pi_2(a_1, a_2) \right]. \end{aligned}$$

Of course, if $n$ is large enough, $\Delta(n - 1) > \eta$, which implies that $\phi \setminus_i \hat{\phi}$ cannot belong to a minimal set satisfying both (a) and (b) in Definition 5.

Finally, we must address the uniqueness of an EAS. But this simply follows from the fact that every two strategy profiles satisfying (2.23) can be connected through admissible transitions (2.24)–(2.27). This completes the proof of the theorem. ∎

When there is no conflict of interest among the players, the above result clarifies the powerful role played by communication in allowing for the evolution of co-operative behaviour. Specifically, when a game of common interest is preceded by the possibility of some stylized form of "cheap" communication, the unique evolutionarily stable set (in the sense of Definition 5) ensures that the single efficient outcome will be achieved in all circumstances, provided that the population is large enough. This latter caveat emphasizes again (recall Section 2.7) the potentially interfering role which "spite" may have in blocking otherwise beneficial mutations in contexts (modelled) with a finite population.

There is some related evolutionary literature which, despite being formulated in a quite different theoretical framework, displays some basic ideas very similar to those of Theorem 2. Noted representatives of it are Fudenberg and Maskin (1990) and Binmore and Samuelson (1992). In both cases, players are assumed involved in a repeated game with no discounting. Their strategies must be either finitely complex (in Fudenberg and Maskin (1990)) or their complexity is assumed costly (in Binmore and Samuelson (1992), where complexity costs are assumed lexicographically less important than the pay-offs strictly derived from playing the game). The key observation to make in this context is that, since no finite history has any effect on intertemporal pay-offs, players can use finite strings of play as a costless mechanism for communication. Relying on such indirect communication, these authors show that any configuration which does *not* satisfy some strong notion of joint rationality (e.g., that does not maximize total pay-offs, in Binmore and Samuelson (1992)) will be destabilized by a "secret handshake" (Robson's felicitous term). Finally, note that the efficiency implications of this approach are, in a sense, much stronger than those of Theorem 2: some efficient outcomes (as, for example, "joint co-operation" in the prisoner's dilemma) may arise as the unique evolutionarily stable configuration, despite not forming part of any "strategic equilibrium" of the stage game.

# 3
# Basic Dynamic Analysis

## 3.1 Introduction

In the previous chapter we made the following "methodological" point. In general, no purely static analysis of evolutionary processes can be judged satisfactory unless complemented with appropriate dynamic foundations. Such a point is, of course, common to many different areas of theoretical research.[1] But, in our case, the need to go beyond static analysis and propose a genuinely dynamic framework represents more than just a methodological consideration. For, as already noted in the preceding chapter, the following two crucial problems with our static approach make it especially unsatisfactory.

First, there is a lack of general conditions under which the central static equilibrium concept, the concept of ESS, can be ensured non-vacuous. For generic games, the reader will recall, it was shown in Section 2.5 that an ESS may *not* exist.

Secondly, the concept of ESS makes good theoretical sense only if the situation it represents is monomorphic, i.e. all members of the population are playing the same strategy. In the generalizations of this concept where the population profile has been allowed to be polymorphic (e.g. the EAS concept proposed in Section 2.8), the validity of the approach crucially relies upon the fact that all individuals are taken to adopt the same action, even though they may play strategies with different prescriptions off the equilibrium path. In general, however, we would like to have a general model for stable and genuine polymorphic situations in which different members of the population may be adopting different actions. As explained in the previous chapter (recall our discussion in Section 2.2) a proper theoretical analysis of these situations cannot avoid an *explicit* modelling of the evolutionary dynamics. To take a first step along this path will be our task in this chapter.

---

[1] Paul Samuelson's well known Correspondence Principle is, for instance, an embodiment of this principle for conventional economic analysis.

## 3.2 The Replicator Dynamics

Recall the general framework introduced in Subsection 2.1.1 and assume, for the moment, that the set of possible phenotypes is finite. For simplicity, we shall assume that they coincide with the set of pure strategies $S = \{s_1, s_2, ..., s_m\}$. This restriction will be relaxed in Section 3.6, where the dynamics involving the *whole set* of mixed strategies will be presented and discussed.

The presentation of the evolutionary dynamics will be first conducted in a discrete-time framework where the motivation of it is clearer. Then, we turn to the continuous-time framework, which may be conceived as the limit of its discrete-time counterpart when either period lengths are short or *relative* fitness differences are small (for example, because all strategies enjoy a relatively large and common "basic fitness").

### 3.2.1 The discrete-time case

Let time be measured discretely, $t = 1, 2, ...$ Denote by $\nu(t) \equiv (\nu_i(t))_{i=1,2,...,m}$, $\sum \nu(\cdot) \equiv 1$, the population profile over pure strategies prevailing at $t$. For simplicity, each member of the population is assumed to live only one period, leaving some offspring which inherits the same phenotype as the parent. Reproduction, therefore, is assumed asexual, with each member of every new generation having only one parent.

Naturally, the number of offspring left by each member of the population is taken to depend on the pay-off earned during his (one-period) lifetime. Specifically, we shall adopt a strictly biological approach here and identify the "pay-off" of an individual with the expected *number* of viable[2] offspring he is able to produce. Thus, if an individual of "generation" $t$ plays strategy $s_i$ against a population profile $\nu(t)$, the expected number of offspring he is assumed to produce (with the same phenotype) exactly equals its pay-off $\pi(s_i, \nu(t)) \equiv \pi_i(\nu(t))$.

With the previous interpretation for pay-offs, the phenotypical dynamics induced becomes a matter of *sheer definition*. Normalize the size of the population to one. Then, if the population profile at $t$ is $\nu(t)$ the size of the population at $t + 1$ is given by:

$$\sum_i \nu_i(t)\, \pi_i(\nu(t)),$$

where recall that each $\pi_i(\nu(t))$ has been identified with the number of offspring produced by *each* individual which plays strategy $s_i$. Thus, $\nu_i(t+1)$, the fraction of the population which plays strategy $s_i$ at $t+1$, is obtained as follows:

$$\nu_i(t+1) = \nu_i(t)\, \frac{\pi_i(\nu(t))}{\sum_{j=1}^{m} \nu_j(t)\, \pi_j(\nu(t))}, \qquad i = 1, 2, ...m \qquad (3.1)$$

---

[2] Here, viability is identified with the capacity for survival and reproduction.

or, denoting the average pay-off $\bar{\pi}\left(\nu\left(t\right)\right) \equiv \sum_{j=1}^{m} \nu_{j}(t)\,\pi_{j}\left(\nu(t)\right)$,

$$\nu_{i}\left(t+1\right)-\nu_{i}\left(t\right)=\nu_{i}\left(t\right)\frac{\pi_{i}\left(\nu(t)\right)-\bar{\pi}\left(\nu(t)\right)}{\bar{\pi}\left(\nu(t)\right)}, \tag{3.2}$$

which, for strategies with positive population frequency $\nu_{i}(t)$ becomes:

$$\frac{\Delta\nu_{i}(t)}{\nu_{i}(t)}\equiv\frac{\nu_{i}(t+1)-\nu_{i}(t)}{\nu_{i}(t)}=\frac{\pi_{i}\left(\nu(t)\right)-\bar{\pi}\left(\nu(t)\right)}{\bar{\pi}\left(\nu(t)\right)}. \tag{3.3}$$

The previous expression has the following intuitive interpretation: the share of the population which plays any given strategy changes in proportion to its relative pay-off (i.e. in proportion to its *deviation*, positive or negative, from the average pay-off).

### 3.2.2    *The continuous-time case*

When adjustments in each period are sufficiently small, the discrete-time dynamics can be approximated, in an arbitrarily close fashion, by a corresponding continuous-time version. An advantage of this alternative formulation is its larger analytical tractability. (For example, the conclusions derived from it do not depend on speeds of adjustment, which can play instead a crucial role when time is modelled discretely.) Moreover, one could argue that, in fact, time is (and therefore should be modelled as) a continuous variable. From this point of view, the division of time in discrete periods is merely an artefact.

To make precise a sense in which one can speak of the above-mentioned approximation, consider a discrete-time framework with variable period length $\Delta > 0$. Following Cabrales and Sobel (1992), assume that, in every period of length $\Delta$, only a fraction $\Delta$ of the population is replaced by its offspring, the rest of the population remaining unchanged. If the fraction of the population which reproduces every period is an unbiased representation of the whole population, and the rate of reproduction of any phenotype is again identified with its average pay-offs, the resulting law of motion can be written (in analogy with (3.1)) as follows:

$$\nu_{i}(t+\Delta)=\frac{\nu_{i}(t)\Delta\,\pi_{i}\left(\nu(t)\right)+\nu_{i}(t)(1-\Delta)}{\sum_{j=1}^{m}\left[\nu_{j}\left(t\right)\Delta\,\pi_{j}\left(\nu(t)\right)+\nu_{j}(t)(1-\Delta)\right]}$$

or[3]

$$\nu_{i}(t+\Delta)-\nu_{i}(t)=\frac{\nu_{i}(t)\left[\Delta\,\pi_{i}\left(\nu(t)\right)-\sum_{j=1}^{m}\nu_{j}\left(t\right)\Delta\,\pi_{j}\left(\nu(t)\right)\right]}{\sum_{j=1}^{m}\left[\nu_{j}\left(t\right)\Delta\,\pi_{j}\left(\nu(t)\right)\right]+(1-\Delta)}.$$

---

[3] Note that for $\Delta = 1$ one obtains (3.2), i.e. the discrete-time version.

Dividing by $\Delta$, and taking limits as $\Delta \to 0$ (i.e. as the length of the time interval becomes infinitesimal), the above expression becomes:

$$\lim_{\Delta \to 0} \frac{\nu_i(t + \Delta) - \nu_i(t)}{\Delta} \equiv \dot{\nu}_i(t) = \nu_i(t)\left[\pi_i(\nu(t)) - \bar{\pi}(\nu(t))\right]; \quad i = 1, 2, ..., m.$$
(3.4)

The above law of motion is labelled the *Replicator Dynamics* (RD) since (in the terminology first suggested by Dawkins (1982)) it models a process of differential "replication" among a given set of strategies or phenotypes. It is the counterpart of (3.2) for a continuous-time framework. Note that the alternative formulation

$$\dot{\nu}_i(t) = \nu_i(t)\frac{\left[\pi_i(\nu(t)) - \bar{\pi}(\nu(t))\right]}{\bar{\pi}(\nu(t))}; \quad i = 1, 2, ...m,$$
(3.5)

which has a closer formal parallelism with the discrete-time version (3.2) will display the same qualitative properties (e.g. the stability of its rest points) as (3.4). This follows from the simple observation that the term $\bar{\pi}(\nu(t))$ in the denominator of (3.5) is common to all of its $m$ equations and, therefore, only affects the "speed" of adjustment of the system, *not* its trajectories.

The dynamical system given by (3.4) can be conceived as the model of an evolutionary process *directly* taking place in continuous time. Alternatively, as explained, it may be interpreted as an approximation to a discrete-time system for which the magnitude of *differential* fitness *per period* is very small. Having the time-period become infinitesimal is one natural way of achieving such small differential fitness per period. A second possibility in this respect is to assume that pay-offs display some large "basic fitness", relative to which differential magnitudes are very small.

To clarify this latter statement, assume that pay-off functions are rewritten in the following form

$$\pi_i(\nu) = \hat{\pi}_i(\nu) + B, \ i = 1, 2, ..., m,$$

for some basic fitness $B$ (independent of $\nu$) and some appropriate set of functions $(\hat{\pi}_i(\cdot))_{i=1,...,m}$. Consider now the discrete-time RD applied to this set-up and make $B \to \infty$ (fixing the functions $(\hat{\pi}_i(\cdot))_{i=1,...,m}$). It is easy to see that the induced (discrete-time) paths converge to those induced by (3.4).

### 3.2.3 Properties of the Replicator Dynamics

#### 3.2.3.1 Invariance of $\Delta^{m-1}$

The above system of differential equations is given by a vector field on $\Delta^{m-1}$. If it is to be well defined, the domain of this vector field must be invariant under (3.4). In other words, every trajectory which starts in $\Delta^{m-1}$ must remain in it when governed by such a law of motion. It is a straightforward task to confirm it. Simply note that:

$$\sum_{i=1}^{m} \dot{\nu}_i (\cdot) = \sum_{i=1}^{m} \{ \nu_i(\cdot) \ (\pi_i(\nu(\cdot)) - \bar{\pi}(\nu(\cdot))) \} \equiv 0$$

and, therefore, $\sum_{i=1}^{m} \nu_i(t) = 1$ if, and only if, $\sum_{i=1}^{m} \nu_i(0) = 1$.

### 3.2.3.2 *Invariance of the boundaries and interior of* $\Delta^{m-1}$

The above property can be strengthened to the following conclusion:

$$\forall \nu(\cdot), \text{ solution to (3.4), } \forall i = 1, 2, ..., m, \forall t \geq 0, \quad \nu_i(t) > 0 \Leftrightarrow \nu_i(0) > 0.$$
$$(3.6)$$

Verbally, (3.6) expresses the idea that if any strategy is played by the population in positive proportion at the start of the process, it will remain played in positive (perhaps very small) proportion all throughout. And conversely, only a strategy which was played in some positive proportion at the start of the process may be played in positive proportion at any time along it.

This property points to an important conceptual problem of the RD: only initially present behaviour may arise along the process. This property of the RD is a consequence of the fact that it only considers *exact* replication (i.e. offspring inherit the same strategy as their parents). This feature will later be modified (see, for example, Section 3.9) by incorporating the possibility that "mutation" may introduce into the process previously inexistent variability.

### 3.2.3.3 *Additive invariance*

Since only relative differences in pay-offs matter for the RD, its behaviour is unaffected by a *common* additive shift in the pay-off functions. This shift, moreover, may depend on the current population profile of strategies.

To be precise, consider any function $\vartheta : \Delta^{m-1} \to \Re$. If the original set of pay-off functions $(\pi_i)_{i=1,2,...,m}$ is replaced by an alternative set $(\hat{\pi}_i)_{i=1,2,...,m}$ such that $\hat{\pi}_i(\cdot) = \pi_i(\cdot) + \vartheta(\cdot)$, the dynamic behaviour of the system is completely unaffected. Mere inspection of (3.4) will confirm this assertion.

### 3.2.3.4 *The Quotient Replicator Dynamics*

Expression (3.4) determines the law of motion of the different strategy frequencies played in the population. Induced by it, we can derive an *equivalent* law of motion for the *relative* frequencies of any pair of strategies. Simple calculation yields:

$$\frac{\dot{\nu}_i(t)}{\nu_i(t)} - \frac{\dot{\nu}_j(t)}{\nu_j(t)} = (\pi_i(\nu(t)) - \pi_j(\nu(t)))$$

for all $i, j = 1, 2, ..., m$ such that $\nu_i(t)$, $\nu_j(t) > 0$. The above law of motion is mathematically equivalent to (3.4), under the interpretation that strategies with zero frequency experience zero proportional rates of change. It describes the RD in an alternative intuitive way: given any two strategies, the relative weights (frequencies) with which they are played in the population change as dictated exclusively by their pay-off difference.

## 3.3   The ESS and the Replicator Dynamics

In this section we start the analysis of the RD by exploring its relationship to the central static concept of ESS presented and discussed in the previous chapter.

For simplicity, subsequent discussion in this chapter will focus on scenarios where contests are random and pairwise, in terms of a certain bilateral game with pay-off matrix $A = (a_{ij})_{ij=1,2,...,m}$ (cf. Subsection 2.1.2). Its extension to more general contexts is generally easy and thus left to the reader.

### 3.3.1   The implicit dynamics of a monomorphic ESS

There is a simple dynamic motivation implicit in the ESS concept (Definition 1): If the ESS strategy (viewed as being initially adopted by the whole population) is threatened by any small mutation, the fact that the latter does strictly worse than the ESS strategy leads to the *implicit* dynamic conclusion that the mutation will eventually disappear.

Such dynamics can be formally described in terms of a replicator system where two (and only two) strategies are possible: the ESS, say $\sigma^*$, and some mutant strategy, denoted by $\rho$. In this context, the state of the system can be fully described by the proportion $\nu_1 \in [0, 1]$ with which the ESS strategy $\sigma^*$ is played in the population ($\nu_2 = 1 - \nu_1$ represents the complementary frequency with which strategy $\rho$ is played). The (one-dimensional) Replicator Dynamics applied to this case has the following form:

$$
\begin{aligned}
\dot{\nu}_1(t) &= \nu_1(t)\, \sigma^* \cdot A\left(\nu_1(t)\sigma^* + (1 - \nu_1(t))\rho\right) - \\
&\quad \nu_1(t) \left\{ \begin{array}{l} \nu_1(t)\left[\sigma^* \cdot A\left(\nu_1(t)\sigma^* + (1 - \nu_1(t))\rho\right)\right] \\ +(1 - \nu_1(t))\left[\rho \cdot A\left(\nu_1(t)\sigma^* + (1 - \nu_1(t))\rho\right)\right] \end{array} \right\} \\
&= \nu_1(t)\,(1 - \nu_1(t)) \left\{ \begin{array}{l} \left[\sigma^* \cdot A\left(\nu_1(t)\sigma^* + (1 - \nu_1(t))\rho\right)\right] \\ -\left[\rho \cdot A\left(\nu_1(t)\sigma^* + (1 - \nu_1(t))\rho\right)\right] \end{array} \right\}
\end{aligned}
$$

which for small $1 - \nu_1(0)$ yields $\dot{\nu}_1(t) > 0$ since $\sigma^*$ is an ESS (and, therefore, the term in brackets above must be positive). This confirms the *dynamic* local stability of the ESS strategy.

### *3.3.2   ESS conditions and polymorphic stability*

As explained in Chapter 2, the interpretation of the ESS concept depends crucially on two assumptions:

(i)   mutations arise only one at a time;

(ii)   the initial state is monomorphic.

Asssumption (i) may be acceptable if mutations are conceived as very rare phenomena with insignificant probability of arising jointly.[4] Assumption (ii), on the other hand, is much less appealing. It seems a fundamental shortcoming to restrict our consideration to situations in which only a single type of behaviour (i.e. strategy) may exist at equilibrium. Instead, behavioural diversity, with all the balances and complementarities that it may afford, often seems a much more likely and natural outcome of evolution.

Such diversity in "observed behaviour" may arise *monomorphically* if the population plays some genuinely mixed strategy. In this case, the pattern of *realized* behaviour — *not* of inherited behaviour, which is monomorphic — will indeed be polymorphic. However, in many contexts, a polymorphic situation is better described as one where the underlying pattern of *inherited* behaviour — not just of realized behaviour — is itself polymorphic. But then, as our previous discussion in Subsection 2.2.2 emphasized, it is necessary to rely on an explicitly dynamic analysis to study its evolutionary stability. This is precisely what is provided by the RD, whose stability features are presently studied.

### *3.3.2.1   Polymorphic stability (I): sufficiency*

Despite its dichotomic and inherently "static" nature, the ESS concept turns out to have a very useful instrumental value in the stability analysis of polymorphic situations. For, as we shall presently show, any ESS (in pure or mixed strategies) always[5] defines a locally stable state of the Replicator Dynamics over pure strategies. However, as we shall illustrate in the next subsection, this unexpected role of the ESS concept cannot be taken too far: there might well exist stable polymorphic states which *cannot* be associated with any corresponding ESS.

For the sake of completeness, we first present the following standard stability notion.

---

[4] This is not the stand taken e.g. in Chs. 5 and 6 where the analysis relies heavily on the possibility that, despite their relatively low probability, simultaneous mutations may indeed arise and play a crucial role in destabilizing otherwise stationary configurations.

[5] In general scenarios (not necessarily with pairwise contests), Theorem 3 below has to be modified slightly: instead of involving the ESS notion, it applies to a strengthening of it that Bomze and Pötscher (1988) call non-invadability. Essentially, this concept requires that there exists a uniform positive threshold across mutations such that, below it, no mutant invasion is possible. In pairwise random contests this notion coincides with that of ESS.

**Definition 6** *Let $\dot{x} = F(x)$ be a dynamical system in $\Re^k$. The point $x^*$ is an asymptotically stable equilibrium of it if:*
*(i) It is Liapunov stable, i.e. given any neighbourhood $U_1$ of $x^*$ there exists another neighbourhood $U_2$ of $x^*$ such that all trajectories with $x(0) \in U_2$ satisfy $x(t) \in U_1, \forall t \geq 0$.*
*(ii) There exists some neighbourhood $V$ of $x^*$ such that all trajectories starting in $V$ satisfy $x(t) \to x^*$ as $t \to \infty$.*

We can now state the main result of this section.

**Theorem 3 (Hofbauer** *et al.* **(1979))** *Let $\sigma^*$ be an ESS. Then, the state $\nu^* = \sigma^*$ is an asymptotically stable equilibrium of RD.*

**Proof.** Let $\sigma^*$ be an ESS. Then, $\forall \sigma \neq \sigma^*, \exists \bar{\varepsilon} > 0$ *s.t. if* $\varepsilon \leq \bar{\varepsilon}$,

$$\sigma^* \cdot A\left[(1-\varepsilon)\sigma^* + \varepsilon\,\sigma\right] > \sigma \cdot A\left[(1-\varepsilon)\sigma^* + \varepsilon\,\sigma\right] \tag{3.7}$$

We first establish the following Lemma:

**Lemma 1** *Condition (3.7) implies:*[6]

$$\exists\, N_{\sigma^*}, \text{ a neighbourhood of } \sigma^*, \text{ s.t. if } \hat{\sigma} \in N_{\sigma^*}, \hat{\sigma} \neq \sigma^*, \sigma^* \cdot A\hat{\sigma} > \hat{\sigma} \cdot A\hat{\sigma} \tag{3.8}$$

*Proof:* Consider the following two possibilities.
If $\sigma^* \in int\left(\Delta^{m-1}\right)$, then simply choose $N_{\sigma^*} = \Delta^{m-1}$. Since, in this case,

$$\sigma^* \cdot A\sigma^* = \hat{\sigma} \cdot A\sigma^*, \quad \forall \hat{\sigma} \in \Delta^{m-1},$$

it follows from Proposition 1 that

$$\sigma^* \cdot A\hat{\sigma} > \hat{\sigma} \cdot A\hat{\sigma}, \quad \forall \hat{\sigma} \in \Delta^{m-1},$$

as desired.
If $\sigma^* \notin int\left(\Delta^{m-1}\right)$, denote by $bd\left(\Delta^{m-1}\right)$ the boundary of the simplex $\Delta^{m-1}$ (the union of all the $(m-2)$-dimensional facets of $\Delta^{m-1}$) and by $bd_{\sigma^*}\left(\Delta^{m-1}\right)$ the set of those facets in $bd\left(\Delta^{m-1}\right)$ which include $\sigma^*$. Let $H \equiv bd\left(\Delta^{m-1}\right) \setminus int\left(bd_{\sigma^*}(\Delta^{m-1})\right)$, where $int(\cdot)$ here denotes the *relative* interior in $bd\left(\Delta^{m-1}\right)$. Clearly, $H$ is relatively compact in $bd\left(\Delta^{m-1}\right)$, and $\sigma^* \notin H$.
For every $\sigma \in H$, let $\tilde{\varepsilon}(\sigma)$ be the supremum over all $\varepsilon > 0$ such that

$$\sigma^* \cdot A\left[(1-\varepsilon)\sigma^* + \varepsilon\,\sigma\right] > \sigma \cdot A\left[(1-\varepsilon)\sigma^* + \varepsilon\,\sigma\right].$$

---

[6] In fact, Conditions (3.7) and (3.8) can easily be seen to be equivalent, i.e. the implication reciprocal to that of the Lemma also holds.

Since $\sigma^*$ is assumed ESS, $\tilde{\varepsilon}(\sigma) > 0$ for all such $\sigma$. Moreover, it defines a continuous function on the compact set $H$. Therefore

$$\min \{\tilde{\varepsilon}(\sigma) : \sigma \in H\} \equiv \varepsilon^* > 0.$$

It is easy to check that the set

$$\{\hat{\sigma} \in \Delta^{m-1} : \hat{\sigma} = (1 - \varepsilon)\sigma^* + \varepsilon\sigma,\ \sigma \in H,\ 0 \leq \varepsilon < \varepsilon^*\}$$

may be selected as the desired neighbourhood $N_{\sigma^*}$. This completes the proof of the Lemma.

To show now the stability of the state $\nu^* = \sigma^*$, define the local Liapunov function $\psi$ on $\Delta^{m-1}$ (see the Appendix):

$$\psi\left(\nu(t)\right) = \prod_{\substack{i=1 \\ \nu_i^* \neq 0}}^{m} \nu_i(t)^{\nu_i^*}.$$

Denote $\vartheta(\cdot) \equiv \log \psi(\cdot)$. For $\nu$ in a sufficiently small neighbourhood of $\nu^*$, say $W$, $\nu_i > 0$ if $\nu_i^* > 0$. Therefore, $\psi(\cdot)$ is positive and $\vartheta(\cdot)$ is well defined in such a neighbourhood. Consider now some other neighbourhood of $\nu^*$ of the form

$$U^* = \{\nu \in \Delta^{m-1} : \vartheta(\nu) \geq \delta\}$$

for some suitable $\delta$. Given any arbitrary neighbourhood $U_1$ (as in Definition 6), choose $\delta$ now large enough such that $U^* \subseteq W \cap U_1 \cap N_{\sigma^*}$ where $N_{\sigma^*}$ is as in the above Lemma. For any $t$ with $\nu(t) \in U^*$ we have:[7]

$$
\begin{aligned}
\frac{\dot{\psi}(\nu(t))}{\psi(\nu(t))} &= \sum_{\substack{i=1 \\ \nu_i^* \neq 0}}^{m} \left[\nu_i^* \frac{\dot{\nu}_i(t)}{\nu_i(t)}\right] = \sum_{i=1}^{m} \{\nu_i^* \left[A_i\nu(t) - \nu(t) \cdot A\nu(t)\right]\} \\
&= \nu^* \cdot A\nu(t) - \nu(t) \cdot A\nu(t) > 0,
\end{aligned}
$$

since $\sigma^*(= \nu^*)$ is an ESS. Making $U_2$ and $V$ in Definition 6 both equal to $U^*$, the desired conclusion follows, completing the proof of the theorem. ∎

### 3.3.2.2    *Polymorphic stability (II): necessity*

The previous result indicates that if the frequencies defining a certain state may be identified with the corresponding probability weights of an ESS, then this is a *sufficient condition* for the state to be asymptotically stable with respect to the RD. An example borrowed from Taylor and Jonker (1978) is now presented, which shows that such a condition is not *necessary* for asymptotic stability.

Consider a scenario with pairwise contests where every two randomly paired individuals play the game defined by the following pay-off table:

---

[7] The notation $A_i$ stands for the $i$th row of the matrix $A$.

|   | $\alpha$ | $\beta$ | $\gamma$ |
|---|----------|---------|----------|
| $\alpha$ | $0,0$ | $1,-2$ | $1,1$ |
| $\beta$ | $-2,1$ | $0,0$ | $4,1$ |
| $\gamma$ | $1,1$ | $1,4$ | $0,0$ |

Table 4

It is straightforward to check that $\tilde{\sigma} = (1/3, 1/3, 1/3)$ is the only symmetric Nash equilibrium of this game. This mixed strategy, therefore, is the only *possible* ESS.

But $\tilde{\sigma}$ is *not* an ESS. To see this, we now show that $\hat{\sigma} = (0, 1/2, 1/2)$ can invade $\tilde{\sigma}$. Making

$$A \equiv \begin{bmatrix} 0 & 1 & 1 \\ -2 & 0 & 4 \\ 1 & 1 & 0 \end{bmatrix}$$

we have:

$$\hat{\sigma} \cdot A\,\tilde{\sigma} = \tilde{\sigma} \cdot A\,\tilde{\sigma} = 2/3;$$
$$\hat{\sigma} \cdot A\,\hat{\sigma} = 5/4 > \tilde{\sigma} \cdot A\,\hat{\sigma} = 7/6,$$

which means that a small fraction of mutants playing $\hat{\sigma}$ may invade a monomorphic population playing $\tilde{\sigma}$.

Consider now the stability of $\tilde{\sigma}$ according to the RD. Denoting $F_i(\sigma) \equiv \sigma_i [A_i \sigma - \sigma \cdot A\sigma]$, the RD is defined by the vector field $F \equiv (F_i)_{i=1,2,3}$. Thus, the local stability of $\tilde{\sigma}$ depends on the Jacobian $DF(\cdot)$ evaluated at $\tilde{\sigma}$. We compute:

$$DF(\tilde{\sigma}) = \frac{1}{9} \begin{bmatrix} -1 & -1 & -4 \\ -7 & -4 & 5 \\ 2 & -1 & -7 \end{bmatrix}.$$

It can be checked that the eigenvalues of the matrix $DF(\tilde{\sigma})$ are $\{-1/3, -1/3, -2/3\}$, all negative real numbers. We may conclude, therefore, that $\tilde{\sigma}$ is locally (asymptotically) stable in terms of the RD even though it is *not* an ESS.

## 3.4    Evolutionary Dynamics and Nash Refinements

The discussion carried out here is parallel to that of Section 2.4. Now, however, the links between equilibrium concepts of Classical and Evolutionary Game Theory will be discussed within a dynamic framework.

The most "primitive" concept to be derived from the RD is that of a stationary state. Specifically, a certain $\nu^* \in \Delta^{m-1}$ is called a *stationary state* if it satisfies:

$$\nu_i^* \left( A_i \nu^* - \nu^* \cdot A\nu^* \right) = 0, \quad i = 1, 2, ..., m, \tag{3.9}$$

where recall that $A_i$ denotes the $i$th row of $A$. The relationship between this concept and that of Nash equilibrium (always applied to the underlying bilateral game with pay-off matrix $A$) is contained in the following proposition.

**Proposition 6** *If $\nu^*$ defines a symmetric Nash equilibrium, it is a stationary state of the RD. The converse need not be true.*

**Proof.** If $(\nu^*, \nu^*)$ is a Nash equilibrium, it is clear that:

$$A_i \nu^* = A_j \nu^* = \nu^* \cdot A\nu^* \tag{3.10}$$

for all $i, j$ such that $\nu_i^* > 0, \nu_j^* > 0$. Thus, (3.9) obviously follows.

Consider, on the other hand, a population state

$$\hat{\nu} = (0, ..., 0, 1, 0, ..., 0)$$

fully concentrated in some strategy $s_j$. Any such vector $\hat{\nu}$ will always satisfy (3.9) trivially, but, in general, it need not define a Nash equilibrium. ∎

The second part of the previous result indicates that, in general, some stationary states will not define Nash equilibria. Clearly, this can only occur if the stationary state in question is *not* interior to $\Delta^{m-1}$. For, otherwise, if an interior point satisfies (3.10), it must also define a Nash equilibrium.

More interestingly, the next result shows that if a stationary state is required to be locally stable, then it must also define a Nash equilibrium.

**Proposition 7 (Bomze (1986))** *Let $\nu^*$ be an asymptotically stable equilibrium of (3.4). Then, $(\nu^*, \nu^*)$ is a Nash equilibrium.*

**Proof.** Let $\nu^*$ be an asymptotically stable point of (3.4), and let $N_{\nu^*}$ be a sufficiently small neighbourhood of it. Take $\nu(0) \in N_{\nu^*}$ such that $\nu_i(0) > 0$ for all $i = 1, 2, ..., m$. By the asymptotic stability of $\nu^*$ we know that $\nu(t) \rightarrow \nu^*$.

Suppose $\nu^*$ were not a Nash equilibrium. Then, there exists some $i, j \in \{1, 2, ..., m\}$ such that $\nu_j^* > 0$ and $A_i \nu^* > A_j \nu^*$. Since $\nu_i(t) > 0$ for all $t$, this implies (cf. the Quotient Replicator Dynamics of Subsection 3.2.3) that there is some $\gamma > 0$ and $T > 0$ such that if $t \geq T$,

$$\frac{\dot{\nu}_i(t)}{\nu_i(t)} - \frac{\dot{\nu}_j(t)}{\nu_j(t)} > \gamma.$$

This obviously yields a contradiction with the assumed convergence to $\nu^*$. ∎

By a more elaborate argument, we can even strengthen the previous theorem as follows.[8]

**Theorem 4 (Bomze (1986))** *Let $\nu^*$ be an asymptotically stable equilibrium of (3.4). Then $(\nu^*, \nu^*)$ is a perfect equilibrium.*

**Proof.** Let $\nu^*$ be an asymptotically stable equilibrium and suppose, for the sake of contradiction, that $(\nu^*, \nu^*)$ is not perfect. By the characterization result already invoked in the proof of Proposition 2, the (mixed) strategy $\sigma^* = \nu^*$ must be dominated by some other strategy $\hat{\sigma} (= \hat{\nu})$. In particular,

$$\hat{\nu} \, A\rho \geq \nu^* \, A\rho, \; \forall \rho \in \Delta^{m-1}.$$

Define $\phi : \Delta^{m-1} \to \Re$ by:

$$\phi(\rho) \equiv \prod_{\substack{i=1 \\ \hat{\nu}_i \neq \nu_i^*}}^{m} \rho_i^{(\hat{\nu}_i - \nu_i^*)}$$

which guarantees that $\phi(\rho) > 0$ for all $\rho >> 0$.[9] For such $\rho$ we can compute:

$$\frac{\partial \phi}{\partial \rho_i}(\rho) = \frac{\hat{\nu}_i - \nu_i^*}{\rho_i} \phi(\rho).$$

We now show that $\nu^*$ is not a local maximum of $\phi$. This is certainly the case if, for some $i \in \{1, 2, ..., m\}$, $\nu_i^* = 0$ but $\hat{\nu}_i \neq 0$. For in this case $\phi(\nu^*) = 0$. Suppose then that for all $i = 1, 2, ..., m$ such that $\nu_i^* = 0$, we have $\hat{\nu}_i = 0$. In this case, $\phi(\nu^*) > 0$ and for some $i'$ for which $\hat{\nu}_{i'} > \nu_{i'}^*$ (one such co-ordinate must exist since $\hat{\nu} \neq \nu^*$ and both belong to the simplex $\Delta^{m-1}$) we have:

$$\frac{\partial \phi}{\partial \rho_i}(\nu^*) > 0. \tag{3.11}$$

Thus, in the direction in which the $i'$ co-ordinate increases so does $\phi(\cdot)$ in a neighbourhood of $\nu^*$. Let $U$ be such a neighbourhood and choose some $\rho^o \in U$ with $\rho_i^o > 0$ for all $i = 1, 2, ..., m$ and $\phi(\rho^o) > \phi(\nu^*)$. This choice is feasible from (3.11).

Consider then a trajectory $\rho(\cdot)$ of (3.4) with $\rho(0) = \rho^o$. The time-derivative of $\phi(\cdot)$ along this trajectory is:

$$
\begin{aligned}
\dot{\phi}(\rho(t)) &= \sum_{i=1}^{m} \frac{\partial \phi}{\partial \rho_i}(\rho(t)) \, \dot{\rho}_i(t) = \sum_{i=1}^{m} (\hat{\nu}_i - \nu_i^*) \frac{\dot{\rho}_i(t)}{\rho_i(t)} \phi(\rho(t)) \\
&= \sum_{i=1}^{m} (\hat{\nu}_i - \nu_i^*) \left( A_i \rho(t) - \rho(t) \cdot A\rho(t) \right) \phi(\rho(t)) \\
&= \left( \hat{\nu} \cdot A\rho(t) - \nu^* \cdot A\rho(t) \right) \phi(\rho(t)) \geq 0,
\end{aligned}
$$

---

[8] Obviously, Theorem 4 makes Proposition 7 redundant. However, for pedagogical reasons, both are included.

[9] The notation ">>" indicates that the strict inequality applies to all vector components.

which implies that $\phi(\cdot)$ does not decrease along any trajectory of (3.4). If such a trajectory starts at $\rho^o$, $\phi(\rho(t)) \geq \phi(\rho^o) > \phi(\nu^*)$. Thus, it cannot converge to $\nu^*$, as claimed. This completes the proof. ∎

**Remark 1** *Note that the preceding theorem, together with Theorem 3, provides an alternative way of proving Proposition 2. By the previous theorem, any asymptotically stable state defines a perfect equilibrium. Furthermore, by Theorem 3, every ESS induces an asymptotically stable state. Thus, every ESS must also be a perfect equilibrium, which is the statement of Proposition 2.*

## 3.5 Some Examples

In order to illustrate the operation of the Replicator Dynamics, two already familiar examples are now reconsidered: the Hawk–Dove and the Rock-Scissors-Papers games, both played in a context with random pairwise meetings.

### 3.5.1 *The Hawk–Dove game revisited (II)*

Consider the Hawk–Dove game, as described in Subsection 2.3.1. We recall its pay-off table:

|  | $H$ | $D$ |
|---|---|---|
| $H$ | $\frac{V-C}{2}, \frac{V-C}{2}$ | $V, 0$ |
| $D$ | $0, V$ | $\frac{V}{2}, \frac{V}{2}$ |

Table 5

which defines the pay-off matrix

$$A = \begin{bmatrix} (V-C)/2 & V \\ 0 & V/2 \end{bmatrix}.$$

If only pure strategies are allowed, the Replicator Dynamics of this game is given by:

$$\dot{\nu}_i(t) = \nu_i(t) \left[ A_i \nu(t) - \nu(t) \cdot A\nu(t) \right]; \quad i = 1, 2.$$

By defining $\nu_1 \equiv x$ and $\nu_2 \equiv 1 - x$, it can be analysed through the following one-dimensional dynamical system:

$$\dot{x}(t) = x(t)\,(1 - x(t))\,[A_1\nu(t) - A_2\nu(t)]\,,$$

or

$$\dot{x}(t) = \frac{1}{2}x(t)\,(1 - x(t))\,[V - Cx(t)]\,. \tag{3.12}$$

We now show that, if $V \leq C$, the equilibrium of this dynamical system, $x = V/C$, is *globally* stable. To verify this, it is enough to find a global Liapunov function.[10] As such a function, we propose the same $\psi(\cdot)$ which was used in the proof of Theorem 3, as particularized to (3.12). That is,

$$\psi\,(\nu(t)) \equiv x(t)^{V/C}\,(1 - x(t))^{(1-V/C)}\,.$$

Along any trajectory of (3.12) we compute:

$$\begin{aligned}
\frac{\dot{\psi}(\nu(t))}{\psi(\nu(t))} &= \frac{1}{2}\frac{V}{C}\,(1 - x(t))\,(V - Cx(t)) + \frac{1}{2}\left(1 - \frac{V}{C}\right)x(t)\,(Cx(t) - V) \\
&= \frac{1}{2C}\,(V - Cx(t))^2
\end{aligned}$$

which, being positive except at the maximum of $\psi(\cdot)$, the point $\left(\frac{V}{C}, 1 - \frac{V}{C}\right)$, proves the desired conclusion. Note that this conclusion does not follow directly from Theorem 3, since this result only establishes *local* stability.

### 3.5.2    The Rock-Scissors-Paper game

Consider the Rock-Scissors-Paper game (R-S-P), described in Section 2.5 for $\varphi = 0$. The pay-offs are summarized by the following table:

|       | R     | S     | P     |
|-------|-------|-------|-------|
| R     | 0, 0  | 1, −1 | −1, 1 |
| S     | −1, 1 | 0, 0  | 1, −1 |
| P     | 1, −1 | −1, 1 | 0, 0  |

Table 6

[10] Of course, a more direct argument for this case could build upon the simple observation that $V - Cx$ (the last term of (3.12)) defines a monotonically decreasing function of $x$ in $[0, 1]$ with a zero in $x^* = V/C$. We resort to a Liapunov function here in order to illustrate the potential *global* use of the approach pursued in the proof of Theorem 3.

which defines the pay-off matrix:

$$A = \begin{bmatrix} 0 & 1 & -1 \\ -1 & 0 & 1 \\ 1 & -1 & 0 \end{bmatrix}.$$

A slightly more complicated argument than the one used in Section 2.5 shows that the R-S-P game allows for no ESS.[11] We lose, therefore, a natural candidate for an asymptotically stable equilibrium. In fact, no such equilibrium exists. For, as we shall presently show, the product $\xi(t) \equiv \nu_1(t)\nu_2(t)\nu_3(t)$ is a *constant of motion*, i.e. a function which remains constant along any trajectory. This indicates that all trajectories are cycles around the equilibrium (1/3,1/3,1/3), each one of them coinciding with a particular level set, $\{\nu \in \Delta^2 : \nu_1 \cdot \nu_2 \cdot \nu_3 = K\}$, for some $0 \leq K \leq 1/27$.

The previous assertion is an immediate consequence of the fact that, along any trajectory of the corresponding RD, $\dot{\zeta}(\cdot) \equiv 0$. Indeed, we compute:

$$\begin{aligned} \dot{\zeta}(t) = & \ \nu_1(t)\nu_2(t)\nu_3(t) \left[\nu_2(t) - \nu_3(t) - 0\right] \\ & +\nu_1(t)\nu_2(t)\nu_3(t) \left[\nu_3(t) - \nu_1(t) - 0\right] \\ & +\nu_1(t)\nu_2(t)\nu_3(t) \left[\nu_1(t) - \nu_2(t) - 0\right] \end{aligned}$$

which obviously equals zero.

The situation is represented in Figure 1.

## 3.6   Replicator Dynamics in Mixed Strategies

### 3.6.1   The model

Suppose now that, in contrast with the dynamic model described in Section 3.2, individuals can adopt *any* of the *mixed* strategies in the set $\Delta^{m-1}$. Then, a state of the system at any given point in time is given by a probability measure $\lambda_t \in \Delta\left(\Delta^{m-1}\right)$, which specifies the different weight with which each mixed strategy $\sigma \in \Delta^{m-1}$ is adopted in the population at $t$.

Given the probability measure

$$\lambda_t : \mathcal{F} \to \Re_+,$$

where $\mathcal{F}$ is a corresponding $\sigma$-field on $\Delta^{m-1}$, the expected pay-off of any strategy $\sigma \in \Delta^{m-1}$ is given by:

$$\pi_t(\sigma) \equiv \pi\left(\sigma, \nu(\lambda_t)\right) = \sigma \cdot A\nu(\lambda_t), \tag{3.13}$$

---

[11] Now it is not possible to argue that there is a mutation that fares strictly better than the unique Nash equilibrium. Rather, the argument is that such a Nash equilibrium does not do *strictly* better than every possible mutation. This is enough to contradict the ESS definition.

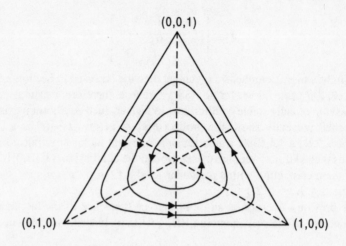

Figure 1: Rock-Scissors-Paper Game, Replicator Dynamics

where, for notational simplicity, we shall make $\nu_t \equiv \nu(\lambda_t)$.

For the moment, time will be considered discrete, $t = 1, 2, \ldots$ From the interpretation of pay-offs explained in Section 3.2, we may write the following law of motion for the frequency of any given mixed strategy $\sigma \in \Delta^{m-1}$:

$$d\lambda_{t+1}(\sigma) = \alpha_t \pi_t(\sigma) d\lambda_t(\sigma), \qquad (3.14)$$

i.e. the frequency (or "density") of individuals adopting strategy $\sigma$ is proportional to their preceding frequency and its former pay-off (cf. the analogous expression (3.1) above). Here, $\alpha_t$ is some appropriate proportionality variable chosen to guarantee that:

$$\int d\lambda_{t+1}(\sigma) = 1.$$

That is:

$$\int \alpha_t \pi_t(\sigma) \; d\lambda_t(\sigma) = \alpha_t \int \pi_t(\sigma) \; d\lambda_t(\sigma) = \alpha_t \left( \int \sigma \; d\lambda_t(\sigma) \right) \cdot A\nu_t = 1$$

which implies:

$$\alpha_t = \frac{1}{\nu_t \cdot A\nu_t},$$

again analogous to the normalizing factor specified by (3.1).

Since the scenario considered involves a population game, only population frequencies over pure strategies matter for determining the individuals' pay-offs.

Motivated by this fact (see also Remark 2 below), our main concern is with the dynamics induced on population states $\nu_t$ as given by:

$$\nu_{i,t+1} = \int \sigma_i \, d\lambda_{t+1}(\sigma), \quad i = 1, 2, ..., m,$$

or, from (3.14):

$$\nu_{i,t+1} = \alpha_t \int \sigma_i \pi_i(\sigma) \, d\lambda_t(\sigma), \quad i = 1, 2, ..., m,$$

which, applying (3.13), becomes:

$$\nu_{i,t+1} = \alpha_t \int \sigma_i(\sigma \cdot A\nu_t) \, d\lambda_t(\sigma) = \alpha_t \left( \int \sigma_i \sigma d\lambda_t(\sigma) \right) \cdot A\nu_t. \qquad (3.15)$$

A standard formula in statistics reads as follows:[12]

$$\int \sigma^\top \sigma d\lambda_t(\sigma) = (\nu_t)^\top \nu_t + \Xi_t, \qquad (3.16)$$

where

$$\Xi_t \equiv \int (\sigma - \nu_t)^\top (\sigma - \nu_t) \, d\lambda_t(\sigma)$$

represents the covariance matrix of the probability measure $\lambda_t$. Denote by $\Xi_{it}$ its $i$th row. Applying (3.16) in (3.15), we obtain:

$$\nu_{i,t+1} = \nu_{i,t} + \alpha_t \, \Xi_{it} \, A\nu_t, \quad i = 1, 2, ..., m. \qquad (3.17)$$

If, as in Subsection 3.2.2, the discrete-time framework is assumed to reflect only small fitness differentials per time-period, (3.17) approximates the following continuous-time system:

$$\dot{\nu}_i(t) = \alpha(t)\Xi_i(t) \, A \, \nu(t), \quad i = 1, 2, ..., m. \qquad (3.18)$$

Because of its tractability, the analysis will focus on this continuous-time version of the dynamics.


### 3.6.2    *ESS conditions and dynamic evolutionary stability*

The objective of this section is to show that, if no a priori restrictions are imposed on the set of possible mixed strategies which can arise in the population, then local stability of any (interior) equilibrium state of (3.18) is *equivalent* to the requirement that this state be an ESS.

---

[12] The notational superindex "T" denotes transpose (here, of row vectors) and is only explicitly used in cases such as this one where dispensing with it may cause confusion. Otherwise, it is avoided for the sake of notational simplicity.

Thus, in this sense, we may say that ESS conditions are *both* necessary and sufficient for evolutionary dynamic stability. This contrasts with the conclusion of Subsection 3.3.2. There, it was shown that if the population can only adopt pure strategies (i.e. population profiles $\lambda$ may only assign positive weight to the "degenerate" mixed strategies identified with the vertices of $\Delta^{m-1}$), the so-restricted RD may have asymptotically stable states which do *not* correspond to any ESS.

A precise statement embodying the previous discussion is contained in the following result.

**Theorem 5 (Hines (1980*b*))** *An equilibrium state $\nu^* \gg 0$ of (3.18) is asymptotically stable if, and only if, $\sigma^* = \nu^*$ is an ESS.*

**Proof.** The asymptotic stability of any interior equilibrium state $\nu^*$ of (3.18) depends on the properties of the matrices $(\Xi(\cdot)A)$ "around" this point. Locally, $\Xi(\cdot)$ can be assumed constant, say equal to $\Xi^*$.[13] Thus, $\nu^*$ will be asymptotically stable if, and only if, the eigenvalues of $\Xi^*A$ all have negative real parts.

However, if no particular restrictions on the strategy profile $\lambda$ around $\nu^*$ are to be considered, *any* positive-definite and symmetric matrix has to be accepted as a possible candidate for $\Xi^*$. This uniform stability of $\nu^*$ then hinges upon whether the eigenvalues of *every* such matrix $\Xi^*A$ have negative real part. In this respect, the following result is shown in the Appendix.

**Lemma 2** *Let $M$ be any square matrix of dimension $m$. The following two statements are equivalent.*
*(i) For every positive definite and symmetric square matrix $Q$, $QM$ has all its eigenvalues with negative real parts;*
*(ii) $(M + M^\top)$ is negative definite.*

Let the pay-off matrix $A$ play the role of the matrix $M$ in the above Lemma. If it is required that $(\Xi^*A)$ have eigenvalues with negative real parts for all positive definite and symmetric $\Xi^*$, this Lemma indicates that it is both necessary and sufficient that $(A + A^\top)$ be negative definite.[14]

Since $\nu^*$ is an equilibrium of (3.18), it is also a Nash equilibrium of the game with pay-off matrix $A$. Therefore, by virtue of Proposition 4, we have that, if $(A + A^\top)$ is negative definite, then $\nu^*$ is an ESS. This completes the necessity part of the Theorem.

As for its sufficiency part, recall that, by Proposition 5, if $\sigma^* = \nu^*$ is an ESS, then $(A + A^\top)$ must be negative definite. In view of the above Lemma,

---

[13] Note that, close to any equilibrium population state $\nu^*$, the differential fitness across pure (and therefore mixed) strategies can be made arbitrarily small. Therefore, the same applies to changes on the covariance matrix.

[14] The sufficiency follows from a standard result in the Theory of Dynamical Systems (see e.g. Hirsch and Smale (1974)). On the other hand, the necessity is implicitly restricted to so-called hyperbolic equilibria (i.e. equilibria whose corresponding eigenvalues do not vanish). Only these equilibria are structurally stable, in the sense of being robust to arbitrarily small perturbations of the functions involved.

the asymptotic stability of $\nu^*$ then follows. This completes the proof of the theorem. ■

**Remark 2** *Note that Theorem 5 does not imply that an asymptotically stable equilibrium $\nu^*$ of (3.18) will induce a corresponding asymptotically stable equilibrium in the strategy profile $\lambda$. Indeed, given any stationary state of (3.18), there is a continuum of alternative population strategy profiles which induce this state. Since this set is connected, it is clearly futile to try to achieve a stringent concept of evolutionary stability in strategy profiles for population games. Only some suitable notion of Liapunov stability (cf. part (i) of Definition 6) can be hoped for in this case.*

## 3.7 Permanence and Survival

The former sections have primarily focused on issues of dynamic stability. It was shown, however, that it is only in cases where an ESS exists that such stability can be guaranteed. And then, of course, only *local* stability around an ESS may be ensured. Nothing has been established, in general, about the global properties of the RD sufficiently far from stationary profiles.

These considerations make it important to have some understanding of the global properties of the RD, even when the latter yield non-convergent paths. In this respect, an often crucial issue turns out to be that of so-called "permanence": essentially, the question of whether pre-existing strategies can be ensured long-run survival. As will be explained, closely connected to this idea is the possibility that well-defined average behaviour will arise along (possibly non-convergent) paths.

### 3.7.1 Definitions

The first and weakest notion of long-run survival which will be considered here is that of *persistence*. Verbally, a certain RD is said to be persistent if, for each of the strategies initially present at the start of the process, its frequency along the ensuing path *cannot* be bounded above at progressively lower levels converging to zero. Persistency, in other words, requires that there should be some positive level for each pre-existing strategy such that, no matter how far into the future the process is being considered, the frequency of this strategy at *some further* date must not fall below that level. Formally:

**Definition 7** *A given RD is said to be persistent if every path $\nu(\cdot)$ with $\nu_i(0) > 0$ for some $i = 1, 2, ..., m$, satisfies $\limsup_{t \longrightarrow \infty} \nu_i(t) > 0$.*

More stringent is the notion of *strong persistence*. It requires that *all* limit frequencies (not just some of them) be positive for all pre-existing strategies.

**Definition 8** *A given RD is said to be strongly persistent if any path $\nu(\cdot)$ with $\nu_i(0) > 0$ for some $i = 1, 2, ..., m$, satisfies $\liminf\limits_{t \longrightarrow \infty} \nu_i(t) > 0$.* '

Finally, the even stronger notion of *permanence* will be of interest for some purposes. Roughly speaking, it demands that the boundary of the state space (the simplex $\triangle^{m-1}$) is a "repeller" of the dynamical system. (That is, any trajectory that approaches the boundary of the simplex from its interior must be "pushed inwards".)

**Definition 9** *A given RD is said to be permanent if there exists some $\delta > 0$ such that all paths $\nu(\cdot)$ with $\nu_i(0) > 0$ for some $i = 1, 2, ..., m$, satisfy $\liminf\limits_{t \longrightarrow \infty} \nu_i(t) > \delta$.*

Notice that strong persistence may be identified with the notion of permanence for $\delta = 0$. Also note that, in this latter concept of permanence, the lower bound $\delta$ is taken uniformly over *all* paths of the dynamical system. It is, therefore, quite a strong notion. Sufficient conditions for it are presented in Subsection 3.7.3. Before this is done, our discussion in the next subsection centres on necessary conditions.

### 3.7.2    *Necessary conditions for persistence and permanence*

As established by the next result, a necessary condition for strong persistence is that the system have some interior rest point, i.e. a stationary state where all pure strategies are played with positive frequency.

**Theorem 6 (Hutson and Moran (1982))** *If an RD is strongly persistent, it has some rest point $\nu \in int(\triangle^{m-1})$.*

**Proof.** By the additive invariance of the RD (see Subsection 3.2.3 above), the dynamics is unaffected if, instead of the original pay-off matrix $A$, an alternative matrix $A'$ is considered which is derived from the former by adding a certain constant to each column. In this fashion, a matrix $A'$ may be constructed so that one of its rows, say the last one, has all its entries equal to zero.

In view of the previous considerations, assume, without loss of generality, that the pay-off matrix $A$ has all of its entries in the $m$th row equal to zero. Then, if $\nu(t) \in int(\triangle^{m-1})$, the quotient RD discussed in Subsection 3.2.3 yields the following law of motion:

$$\frac{\dot{\nu}_i(t)}{\nu_i(t)} - \frac{\dot{\nu}_m(t)}{\nu_m(t)} = A_i\nu(t) - A_m\nu(t) = A_i\nu(t), \quad i = 1, 2, ..., m.$$

Denoting $\xi_i(t) \equiv \frac{\nu_i(t)}{\nu_m(t)}$ we may rewrite the former expression as follows:

$$\dot{\xi}_i(t) = (\xi_i(t) A_i \xi(t)) \, \nu_m(t), \quad i = 1, 2, ..., m, \tag{3.19}$$

where, for formal simplicity in what follows, we include the $m$th trivial component for which $\xi_m(\cdot) \equiv 1$ and $\dot{\xi}_m(\cdot) \equiv 0$.

Obviously, there is a one-to-one correspondence between the interior limit points of the RD (as well as, a *fortiori*, its interior rest points) and those of (3.19). Moreover, since all equations in (3.19) are multiplied by a common factor $\nu_m(t) \geq \eta > 0$ (by strong persistence), the same applies when, instead of (3.19), we consider the following dynamical system:

$$\dot{\xi}_i(t) = \xi_i(t) A_i \xi(t), \quad i = 1, 2, ..., m. \tag{3.20}$$

Thus, assume, for the sake of contradiction, that (3.20) has no interior rest points. Then, denoting by $\Re_{++}^{m-1}$ the interior of the positive orthant of $\Re^{m-1}$ and $X \equiv \Re_{++}^{m-1} \times \{1\}$, the set

$$W = \{ \, y \in \Re^m : y = A\xi, \, \xi \in X \}$$

does not include the origin $0 \in \Re^m$. Since $W$ is obviously convex, the well-known Separation Theorem of Convex Analysis implies that there exists some vector $\mu \in \Re^m$, $\mu \neq 0$, which separates $W$ from 0, i.e.

$$\forall y \in W, \quad \mu \cdot y > \mu \cdot 0 = 0. \tag{3.21}$$

Consider now the function $\varphi : X \to \Re$ given by $\varphi(\xi) = \sum_{i=1}^m \mu_i \log \xi_i$. The time derivative of $\varphi(\cdot)$ along any path of the system in the interior of $X$ is given by:

$$\dot{\varphi}(\xi(t)) = \sum_{i=1}^m \mu_i \, A_i \, \xi(t), \tag{3.22}$$

which is strictly positive, by (3.21). This, as we shall now argue, implies that no limit point of the dynamics (3.20) can lie in $\Re_{++}^m$. By previous considerations, the same happens with the limit points of (3.19). But then, because of the correspondence between the interior limit points of (3.19) and the RD, it follows that no limit point of the RD can lie in $int(\triangle^{m-1})$, contradicting the assumed strong persistence of the system.

Thus, to complete the proof of the theorem, it needs to be shown that there *cannot* exist some $\xi^0 \in \Re_{++}^m$ which is a limit point of (3.20). Suppose otherwise. Then, for any such $\xi^0$, there must be a sequence of "times" $(t_k)_{k=1,2...}$, such that:

$$\lim_{k \to \infty} \xi(t_k) = \xi^0. \tag{3.23}$$

By the continuity of both $\varphi(\cdot)$ and $\dot{\varphi}(\cdot)$, (3.23) implies that, given any $\varepsilon_1 > 0$, $\varepsilon_2 > 0$, there is some $K \in \mathcal{N}$ such that, if $k_0 \geq K$,

$$\left| \varphi(\xi(t_{k_0})) - \varphi(\xi^0) \right| \leq \varepsilon_1 \,,$$

$$\left| \dot{\varphi}(\xi(t_{k_0})) - \dot{\varphi}(\xi^0) \right| \leq \varepsilon_2 \,,$$

which, in view of (3.21) and (3.22), implies that, for some $\delta > 0$, $\varphi\left(\xi(t_{k_0} + \delta)\right) > \varphi(\xi^0)$. Since $\dot{\varphi}\left(\xi(t)\right) \geq 0$ for all $t$, this contradicts the fact that $\xi^0$ is a limit point of the dynamics, thus completing the proof of the theorem.   ∎

Since strong persistence is a concept weaker than permanence, the following Corollary follows:

**Corollary 1** *If an RD is permanent, it has some rest point $\nu \in int(\triangle^{m-1})$.*

### 3.7.3    Sufficient conditions for permanence

In order to establish the desired sufficient conditions, the following auxiliary result is needed.

**Theorem 7 (Hofbauer and Sigmund (1987))** *Consider some given RD and let $\varphi : \triangle^{m-1} \to \Re$ be a differentiable function which vanishes in $bd(\triangle^{m-1})$ and is strictly positive elsewhere. If there is a continuous function $\phi : \triangle^{m-1} \to \Re$ which satisfies the following two conditions:*

$$\nu(t) \in int(\triangle^{m-1}), \ t \succeq 0, \ \frac{\dot{\varphi}\left(\nu(t)\right)}{\varphi\left(\nu(t)\right)} = \phi(\nu), \qquad (3.24)$$

$$\forall \nu(t) \in bd(\triangle^{m-1}), \ t \succeq 0, \int\limits_{t}^{T} \phi\left(\nu(t)\right) dt > 0 \text{ for some } T > t, \qquad (3.25)$$

*then the dynamical system is permanent.*

The former result is an intuitive generalization of the so-called Second Liapunov Theorem, applied to the RD. Instead of using Liapunov functions in the usual sense (functions whose time-derivatives are of definite sign along any trajectory), it relies on "average Liapunov functions", i.e. Liapunov functions whose time-integrals over finite horizons have the required definite sign.

Because of its merely instrumental role here, the proof of Theorem 7 is not included. The interested reader is referred to Hofbauer and Sigmund (1988, Theorem 1, p. 98) for a detailed proof of this result. Relying on it, the following Theorem is now established.

**Theorem 8 (Hofbauer and Sigmund (1987))** *Assume that there exists some $\eta \in int(\triangle^{m-1})$ such that*

$$\eta \cdot A\zeta > \zeta \cdot A\zeta \qquad (3.26)$$

*for all rest points $\zeta$ in $bd(\triangle^{m-1})$. Then the RD is permanent.*

**Proof.** With the aim of applying Theorem 7, define $\varphi : (\triangle^{m-1}) \to \Re$ as follows:

$$\varphi(\nu) = \prod_{i=1}^{m} \nu_i{}^{\eta_i}.$$

The function $\varphi(\cdot)$ obviously satisfies the requirements contemplated by Theorem 7. Consider now the function $\phi : (\triangle^{m-1}) \to \Re$ defined by:

$$\phi(\nu) = \eta \cdot A\nu - \nu \cdot A\nu,$$

which clearly fulfils (3.24). Thus, the desired conclusion only requires the verification of (3.25).

Given any $t$ and the corresponding $\nu(t)$, the argument proceeds inductively on the number of components $r = 1, 2, ..., m$ of $\nu(t)$ which are positive. Initially, let $r = 1$. Then, $\nu(t)$ is a vertex of the simplex $\triangle^{m-1}$ and (3.25) immediately follows from (3.26) and the fact that every simplex vertex is stationary (cf. Proposition 6). Assume now that (3.25) holds for all $r = 1, 2, ..., s$, with $s \leq m - 2$. We need to show that it also holds for $r = s + 1$.

Let $\nu(t)$ have all but $s + 1$ of its components vanish. Then the $(s+1)$-dimensional vector formed by the positive components of $\nu(t)$ belongs to the simplex $\triangle^s$. Two possibilities arise:

(i)   $\nu(\tau)$ converges to $bd(\triangle^s)$ as $\tau \to \infty$;

(ii)  $\nu(\tau)$ does not converge to $bd(\triangle^s)$ as $\tau \to \infty$.

If (i) applies, the fulfilment of (3.25) for $r = s$ is easily seen to imply that it must also hold for such a trajectory. (Here, the main point to observe is that, if $\nu(\tau)$ remains in some sufficiently small neighbourhood $bd(\triangle^s)$ after some time $\tau_0$, the subsequent repeated application of (3.25) will offset any prior negative value for the average Liapunov function.)

If (ii) applies, one may choose some $\varepsilon > 0$ and a sequence $\{t_k\}$ of times, $t_k \geq t$, $t_k \to \infty$, such that $\nu_i(t_k) \geq \varepsilon$ for $i = 1, 2, ..., s + 1$ (for simplicity, it is assumed that the non-vanishing components of $\nu(\cdot)$ correspond to those with the first $s+1$ indices). Consider the associated sequences $\{\tilde{\nu}(t_k)\}_{k=1,2,...}$, $\{\rho(t_k)\}_{k=1,2,...}$, defined as follows:

$$\tilde{\nu}(t_k) = \frac{1}{t_k - t} \int_t^{t_k} \nu(\tau) \ d\tau , \qquad (3.27)$$

$$\rho(t_k) = \frac{1}{t_k - t} \int_t^{t_k} \nu(\tau) \cdot A\nu(\tau) \ d\tau . \qquad (3.28)$$

Since both sequences are bounded, there is a (common) subsequence $\{\hat{t}_k\} \subseteq \{t_k\}$ such that the corresponding subsequences $\{\tilde{\nu}(\hat{t}_k)\}_{k=1,2,...}$, $\{\rho(\hat{t}_k)\}_{k=1,2,...}$ are convergent. Let $\tilde{\nu}*$ and $\rho*$ denote their respective limits.

Consider now the sequences $\{\varpi_i(\hat{t}_k)\}_{k=1,2,\dots}$, $i = 1, 2, \dots, s + 1$, defined as follows:

$$\varpi_i(\hat{t}_k) \equiv \frac{1}{t_k - t} \int_t^{\hat{t}_k} \frac{\dot{\nu}_i(\tau)}{\nu_i(\tau)} \, d\tau = A_i \tilde{\nu}(\hat{t}_k) - \rho(\hat{t}_k). \tag{3.29}$$

Since

$$\int_t^{t_k} \frac{\dot{\nu}_i(\tau)}{\nu_i(\tau)} \, d\tau = \log \nu_i(t_k) - \log \nu_i(t)$$

is bounded by the assumption that $\nu_i(t_k) \geq \epsilon$ for all $t_k$, it follows that $\varpi_i(\hat{t}_k) \to 0$ as $t_k \to \infty$. Therefore,

$$A_i \tilde{\nu}* = \rho*, \quad i = 1, 2, \dots, s + 1,$$

which implies that $\tilde{\nu}^*$ is a rest point of the RD. From these considerations, it follows that the sequence $\{\zeta(\hat{t}_k)\}$ defined by:

$$\zeta(\hat{t}_k) \equiv \frac{1}{t_k - t} \int\limits_t^{\hat{t}_k} \phi\left(\tilde{\nu}^*(\tau)\right) d\tau = \eta \cdot A_i \tilde{\nu}(\hat{t}_k) - \rho(\hat{t}_k)$$

converges to $\eta \cdot A_i \tilde{\nu}^* - \rho^*$, which is positive by (3.26). This confirms (3.25) for case (ii) and the proof is complete. ∎

### 3.7.4    Average behaviour in permanent systems

The previous subsections have determined necessary and sufficient conditions for an RD to display permanence. As presently shown, the notion of permanence has, beyond its inherent interest, useful long-run implications. In particular, it ensures that the system will exhibit a well-defined and *unique* average behaviour in the long run.

As established by Theorem 9 below, the long-run frequencies of a permanent system are directly associated with the corresponding frequencies of some interior equilibrium, the existence of which was shown in Theorem 6 to be a necessary condition for permanence. Thus, in order to ensure that long-run behaviour is uniquely defined, the existence of a *unique* interior equilibrium must first be established. This is done by the next result, an immediate strengthening of Theorem 6:

**Proposition 8** *If an RD is permanent, it has a unique rest point $\nu \in int\left(\triangle^{m-1}\right)$.*

**Proof.** By Theorem 6 and its Corollary, every permanent system has some interior rest point. To prove that any such rest point must be unique, suppose a given permanent RD had two of them, $\nu^1$ and $\nu^2$, $\nu^1 \neq \nu^2$.

Since both $\nu^1$ and $\nu^2$ are *interior* rest points of the RD they must satisfy:

$$\forall i, j = 1, 2, ..., m \quad A_i \nu^h = A_j \nu^h, \quad \text{for each} \quad h = 1, 2.$$

Therefore, it is clear that every population state in the set

$$E = \{\nu \in \triangle^{m-1} : \nu = \alpha \nu^1 + (1 - \alpha) \nu^2, \ \alpha \in \Re\}$$

must also be an equilibrium of the RD. This set defines a line in $\triangle^{m-1}$ which obviously intersects $bd\left(\triangle^{m-1}\right)$ at some $\hat{\nu}$. For this state, $\hat{\nu}_{i'} = 0$ for some $i' = 1, 2, ..., m$.

Consider now any $\delta > 0$, as contemplated in Definition 9. Clearly, there is an element $\tilde{\nu}$ in the set $E$, sufficiently close to $\hat{\nu}$, such that $\tilde{\nu} \in int\left(\triangle^{m-1}\right)$ and $\tilde{\nu}_{i'} < \delta$. Since the point $\tilde{\nu}$ is a rest point of the RD, the system cannot be permanent. This contradiction completes the proof of the Theorem. ■

On the basis of the preceding result, the main conclusion of this subsection can now be established: the long-run behaviour of a permanent RD is uniquely defined.

**Theorem 9 (Hofbauer and Sigmund (1987))** *Let* $\nu(\cdot) \subset int\left(\triangle^{m-1}\right)$ *be any path of a permanent RD. It satisfies:*

$$\lim_{T \to \infty} \frac{1}{T} \int_{t=0}^{T} \nu(t) \ dt = \nu^*,$$

*where* $\nu^*$ *is the unique interior equilibrium.*

**Proof.** For $\nu(t) \in int(\triangle^{m-1})$, write the RD as follows:

$$(\log \nu_i(t))^{\cdot} = A_i \nu(t) - \nu(t) \cdot A\nu(t), \quad i = 1, 2, ..., m. \tag{3.30}$$

Given any $t > 0$, integrate (3.30) over the time-interval $[0, t]$ and divide by its length to obtain, for each $i = 1, 2, ..., m$,

$$\frac{\log \nu_i(t) - \log \nu_i(0)}{t} = \frac{1}{t} \int_{\tau=0}^{t} A_i \nu(t) - \frac{1}{t} \int_{\tau=0}^{t} \nu(t) \cdot A\nu(t)$$

$$= A_i \hat{\nu}(t) - \hat{\rho}(t), \tag{3.31}$$

where, by analogy with (3.27) and (3.28) above, $\hat{\nu}(t)$ and $\hat{\rho}(t)$ are defined as follows:

$$\hat{\nu}(t) = \frac{1}{t} \int_{\tau=0}^{t} \nu(\tau) \ d\tau,$$

$$\hat{\rho}(t) = \frac{1}{t} \int\limits_{\tau=0}^{t} \nu(\tau) \cdot A\nu(\tau) \ d\tau.$$

The remainder of the argument is essentially parallel to that of the proof of Theorem 8. Consider a sequence $\{t_k\}$ of times, $t_k \to \infty$, and their corresponding sequences $\{\hat{\rho}(t_k)\}_{k=1,...}$, $\{\hat{\nu}_i(t_k)\}_{k=1,...}$, $i = 1, 2, ..., m$. The sequences $\{\hat{\nu}_i(t_k)\}_{k=1,...}$ are obviously bounded above. Moreover, since the RD is assumed permanent, they are also bounded below by some $\delta > 0$. Thus, there must exist some subsequence $\{\hat{t}_k\}_{k=1,...}$ such that each of the subsequences $\{\hat{\rho}(\hat{t}_k)\}_{k=1,...}$, $\{\hat{\nu}_i(\hat{t}_k)\}_{k=1,...}$ is convergent to some corresponding $\hat{\rho}^*$ and $(\hat{\nu}_i^*)_{i=1,2,...,m}$ with $\hat{\nu}_i^* > 0$ for each $i = 1, 2, ..., m$. Since the LHS of (3.31) converges to zero as $t \to \infty$, it follows that:

$$\forall i = 1, 2, ...m, \quad A_i\hat{\nu}_i^* = \hat{\rho}^* = \hat{\nu}_i^* \cdot A\hat{\nu}_i^*.$$

Therefore, the average frequencies, $(\hat{\nu}_i^*)_{i=1,2,...,m}$ define an interior equilibrium. By Proposition 8, such equilibrium and, therefore, the average frequencies are unique. This completes the proof of the Theorem. ∎

## 3.8   Population Genetics

As emphasized in Subsection 2.1.1, the stylized model of phenotypical evolution given by the RD abstracts from genetic considerations. Phenotypes themselves are taken to be the object of inheritance, a process modelled in a fully asexual manner. Of course, this is not a faithful description of what happens in the biological world, where reproduction is both genetic (i.e. not "phenotypic") and often sexual.

In contrast, population genetics concerns itself with the *genetic* evolution of *sexual* populations, its traditional focus being on models where individual pay-offs are assumed *independent* of population frequencies (i.e. exclusively linked to the genetic make-up of each individual). The underlying emphasis and motivation of classical population genetics is, therefore, polar to that exhibited by the RD. However, as explained below, both contexts exhibit close formal parallelisms. In both of them, the dynamics of relative growth is modelled though replicator equations, only their respective interpretations being, of course, substantially different. A comparison of both approaches (their similar features, but also their qualitatively different conclusions) will shed new light on our previous analysis of the RD.

Consider the simplest case of some interest in population genetics: a *one-locus diploid* genetic system. In this context, the fitness of any given individual depends on just *one* single locus, *two* alleles occupying it to determine the individual's phenotypical (and, therefore, pay-off-relevant) characteristics.[15]

Let $\mathcal{L} = \{a_1, a_2, ..., a_m\}$ be the set of alleles (the "genetic pool") which is a priori available to the population. Since the genetic system is assumed diploid, there are $m^2$ possible genotypes, corresponding to each of the possible pairs (combinations) of alleles $(a_i, a_j)$, $i, j = 1, 2, ..., m$, which an individual may have. The *fitness* of each genotype $(a_i, a_j)$ is captured by the so-called *fitness matrix* $W \equiv (w_{ij})_{i,i=1,2,...,m}$. A typical entry of it, $w_{ij}$, indicates the pay-off obtained by an individual whose genotype is $(a_i, a_j)$. Since genotype $(a_i, a_j)$ is identical to $(a_j, a_i)$, the matrix $W$ must be *symmetric*, i.e. $w_{ij} = w_{ji}$, $\forall i, j = 1, 2, ..., m$. Symmetry of the fitness matrix $W$ is one of the key features of the present context. The other one, as explained, is the assumption that individuals' pay-offs are *exclusively* linked to their genotype (i.e. they are frequency *independent*).

Suppose that reproduction is *sexual* and mating is *random*, with each of the parents of the couple transmitting *one* of its alleles to their offspring with equal probability. For simplicity, assume that every couple produces the same number of offspring, irrespective of their genetic make-up. Not every new offspring, however, has the same chance of becoming a viable adult and thus capable of further reproduction. This depends on its fitness (as given by the matrix $W$), which may be interpreted as (proportional to) its probability of reaching sexual maturity.

More precisely, let $\xi(t) \equiv (\xi_1(t), \xi_2(t), ..., \xi_m(t)) \in \Delta^{m-1}$ denote the population frequencies prevailing at $t$ for each of the $m$ alleles. From our assumption of random sexual mating, the frequency of (not yet necessarily viable) offspring with heterozygotic genotype $(a_i, a_j)$ in the new generation must be $2\xi_i\xi_j$ if $i \neq j$. (Notice that, since genotypes $(a_i, a_j)$ or $(a_j, a_i)$ are equivalent, the same genotype results if the "first" individual contributes allele $a_i$ and the "second" $a_j$, or vice versa.) On the other hand, the frequency of the homozygotic offspring $(a_i, a_i)$ will simply be $\xi_i^2$.

As explained, not all these offspring reach maturity with the same probability. To obtain the genotype frequencies resulting at maturity, we must weigh the above computed frequencies by their respective fitness. Thus, the frequency of new mature individuals with genotype $(a_i, a_j)$ in the next generation is proportional to $2w_{ij}\xi_i\xi_j$. For those individuals whose genotype is $(a_i, a_j)$, it will be proportional to $w_{ii}\xi_i^2$.

From these fitness-weighted frequencies over genotypes in the next generation, we can obtain the new vector of *allele* frequencies ready for fresh combination and reproduction. Adding up across all possible genotypes, we conclude that the frequency of each allele $a_i$ in the next generation involves the following two terms. First, one "copy" of $a_i$ exists in all heterozygotic individuals with genotypes $(a_i, a_j)$ – or $(a_j, a_i)$ – with $i \neq j$. As explained above, the aggregate frequency for these individuals is proportional to $\sum_{j \neq i} 2 w_{ij} \xi_i \xi_j$. On

---

[15] An allele is each of the independent pieces (or "bits") of genetic information specified in every genetic locus.

the other hand, there are *two* copies of allele $a_i$ in every homozygotic individual with genotype $(a_i, a_i)$. The allele frequency in this latter case is therefore proportional to $2\, w_{ii}\, \xi_i^2$.

Denote by $\xi_i'$ the frequency of allele $a_i$ in the next generation. From the previous considerations, we conclude that

$$
\begin{aligned}
\xi_i' &= \alpha \left[ \sum_{j \neq i} 2\, w_{ij}\, \xi_i \xi_j + 2\, w_{ii}\, \xi_i^2 \right] \\
&= 2\alpha\, \xi_i \sum_{j=1}^{m} w_{ij}\, \xi_j,
\end{aligned}
$$

for some common $\alpha > 0$. Thus, $\xi_i'$ is proportional to $\xi_i$ (the preceding frequency) *and* the average pay-off $\sum_{j=1}^{m} w_{ij}\, \xi_j$ associated with allele $a_i$ in the preceding generation, as measured by the average fitness of the individuals which hold it.

Formally, therefore, the dynamics induced are *exactly* as for the RD. If time is modelled continuously (possibly relying on considerations analogous to those explained in Subsection 3.2.2 for the RD), the law of motion for $\xi(t) = (\xi_1(t), \xi_2(t), ..., \xi_m(t))$ is of the following form:

$$
\dot{\xi}_i(t) = \xi_i(t)\, [W_i \xi(t) - \xi(t) \cdot W \xi(t)]; \quad i = 1, 2, ..., m, \tag{3.32}
$$

which is the same as (3.4), with the obvious notational adjustments.

As explained, there is a crucial difference between the RD and the model of genetic dynamics outlined here: in the present context, the pay-off (fitness) matrix $W$ has to be symmetric, since allele permutation yields a fully equivalent genotype. Due to this symmetry, the dynamics of population genetics leads to sharp and strong conclusions which are generally out of reach in the RD context. A classical result of this type, often labelled *Fisher's Fundamental Theorem of Natural Selection*, reads as follows.

**Theorem 10** *The average fitness $\xi(t) \cdot W \xi(t)$ increases monotonically along any trajectory of (3.32).*

**Proof.** It is enough to show that the time-derivative of the function $\phi(t) \equiv \xi(t) \cdot W \xi(t)$ is positive along any trajectory $\xi(\cdot)$ of the dynamical system. We compute:

$$
\dot{\phi}(t) = \dot{\xi}(t) \cdot W \xi(t) + \xi(t) \cdot W \dot{\xi}(t).
$$

Since $W$ is symmetric,

$$
\dot{\xi}(t) \cdot W \xi(t) = \xi(t) \cdot W \dot{\xi}(t),
$$

and therefore:

$$
\dot{\phi}(t) = 2\, \dot{\xi}(t) \cdot W \xi(t)
$$

which, applying (3.32), becomes:

$$\dot{\phi}(t) = 2 \sum_{i=1}^{m} \xi_i(t) [W_i\xi(t) - \xi(t) \cdot W\xi(t)] \; W_i \xi(t)$$

$$= 2 \left\{ \sum_{i=1}^{m} \xi_i(t) [W_i\xi(t)]^2 - \left[ \sum_{i=1}^{m} \xi_i(t) W_i\xi(t) \right]^2 \right\} \quad (3.33)$$

$$= 2 \sum_{i=1}^{m} \xi_i(t) [W_i\xi(t) - \xi(t) \cdot W_i\xi(t)]^2 ,$$

where the last term is simply twice the fitness variance at $t$, obviously positive unless

$$W_i\xi(t) = \xi(t) \cdot W\xi(t) ,$$

for all $i = 1, 2, ..., m$ such that $\xi_i(t) \neq 0$, that is, unless the system is at a stationary point.  ∎

The previous result underscores the important point that, if one abstracts from the "complexities" of frequency-dependent selection (precisely what is at the heart of the RD) natural selection operating on the genetic base of the population always has the unambiguous effect of increasing its average fitness. In a sense, we can view this elegant result as the paradigm of what natural selection could achieve *if* (of course, a major *"if"*) it were to function without the interference on pay-offs induced by intraspecific interaction. Reciprocally, it highlights the important role of frequency-dependent considerations in understanding realistic evolutionary dynamics.

To close this section, we point to two very interesting implications of Theorem 10.

**Remark 3** *Expression (3.33) indicates that the rate of change of (3.32), as measured by the rate at which average fitness changes along any given trajectory, is proportional to the current fitness variance. This reflects the intuitive idea that natural selection thrives on heterogeneity, if not of genes per se at least of pay-offs.*

**Remark 4** *Generically, the stationary points induced by a certain fitness matrix W must be isolated. In combination with Theorem 10, this implies that, generically, all trajectories of the system (3.32) will converge to a limit point (not necessarily the same one for all initial conditions). In fact, this conclusion on long-run convergence may be shown to be true (see, for example, Losert and Akin (1983)) even without the genericity proviso. This, of course, contrasts sharply with the behaviour of the RD, as discussed for example in Section 3.7.*

## 3.9 The Prisoner's Dilemma

The prisoner's dilemma has become a paradigmatic benchmark for the analysis of many interesting game-theoretic issues: bounded rationality, repeated interaction, incomplete information, etc. It should be useful, therefore, to turn towards this traditional context for an illustration of our evolutionary models.

Consider, for concreteness, a prisoner's dilemma (PD) with the following pay-off matrix:

|     | $C$ | $D$ |
| :-: | :-: | :-: |
| $C$ | 3,3 | 0,4 |
| $D$ | 4,0 | 1,1 |

Table 7

As is well known, each player has an unambiguous dominant strategy in a PD ($D$ above, standing for "defection") which leads to an inefficient outcome. In particular, if both players were to choose strategy $C$ ("co-operation"), both would be better off than by jointly playing their dominant strategy $D$.

In the face of this state of affairs (sometimes labelled the "prisoner's dilemma paradox"), the received game-theoretic literature has proposed quite a number of different approaches. A key one has been to postulate that players are involved in repeated interaction, an identical PD being played over time between the same two players. As it turns out, the possibilities open to the players in this repeated context (at least within the classical strategic framework) crucially depend on the duration of the relationship. If finite, simple backwards induction (from the last time at which the game is played to any former period) forces the agents to the same "rational" outcome as before: joint defection throughout. However, if the situation can be appropriately modelled as one with an infinite horizon, a wide range of new possibilities arises. To be sure, one of them still is joint indefinite defection. But, in addition, if players do not discount future pay-offs too much, indefinite co-operation is another possibility. It is built upon the threat that if any player ever deviates, the opponent will "punish" him with defection thereafter. In fact, the so-called *folk theorems* have established the following general result: If players are sufficiently patient, *any* outcome that yields at least the maximin pay-off to each of them (in our case, a pay-off of 3, which both may guarantee for themselves by playing $D$) can be supported as a Nash equilibrium of the infinitely repeated game.

We shall not elaborate upon this well-known literature since it is only of tangential relevance to our present concerns.[16] Partly in response to it, there

---

[16] The interested reader may refer to the classical book by Friedman (1977) for an exhaustive treatment of this subject.

was a lively resurgence of interest in the repeated prisoner's dilemma (RPD) during the 1980s, whose focus was much more evolutionary than strategic. It started with the pioneering work of Axelrod and Hamilton (1981) – see also Axelrod (1984). His concern was not so much to evaluate the "rationality" of any given strategy *per se*, but rather to understand its potential for long-run survival in competition with other alternative strategies.

Axelrod's first exploration of these issues sprang from a series of experiments of "natural selection" conducted on a collection of strategies for the RPD submitted by a number of different game theorists. To the initial surprise of many, the strategy which fared relatively better in the long run was the natural (and simple) one suggested by Anatol Rapoport: tit-for-tat. Since then, a number of different lines of research have underscored the central role played by simple and reactive strategies in the long-run evolution of co-operation.[17]

This fast-growing literature is already too vast and rich to make a short summary possible.[18] Instead, we shall focus on a simplified version of the RPD that illustrates the operation of evolutionary forces (in particular, the Replicator Dynamics) in the rise of co-operative behaviour. As an interesting by-product, we shall be exposed for the first time in this book to an idea which will play a much more central role in future analysis. Very succinctly, it points to the drastic qualitative effects which the introduction of small amounts of noise (due, for example, to mutation) may have on the long-run performance of the evolutionary process.

### 3.9.1    Basic (unperturbed) model

Consider the following theoretical, very stylized, context. Individuals of a large population (with the cardinality of the continuum) are randomly matched every period to play an RPD with an infinite horizon. They adopt one of the following three (repeated-game) strategies:

- Strategy $C^*$, which chooses the action $C$ throughout, irrespective of past history.

- Strategy $D^*$, which chooses the action $D$ throughout, again irrespective of past history.

- Strategy $T$ (tit-for-tat), which starts by co-operating and, subsequently, responds by matching the opponent's action in the preceding stage.

Assume that players' intertemporal pay-off in each matching round equals the discounted sum of their corresponding stream of pay-offs. Suppose, for

---

[17] See e.g. the recent paper by Nowak and Sigmund (1992).

[18] The interested reader may refer to the recent monographic issue of *BioSystems* on this topic, Fogel (1996).

concreteness, that both of them have a common discount rate $\delta = 2/3$. Then, we can set up the following nine-entry table, which summarizes the pay-offs resulting from every possible matching in the RPD under consideration. (As customary, discounted pay-offs are multiplied by the factor $(1 - \delta)$, in order to have them remain within the convex hull of stage pay-offs.)

|       | $C^*$ | $D^*$ | $T$    |
| ----- | ----- | ----- | ------ |
| $C^*$ | $3,3$ | $0,4$ | $3,3$  |
| $D^*$ | $4,0$ | $1,1$ | $2,2/3$ |
| $T$   | $3,3$ | $2/3,2$ | $3,3$ |

Table 8

Consider now the Replicator Dynamics applied to this context on the basis of the expected (or average) pay-offs earned by each strategy. Denote by $x(t)$ the frequency of $D^*$-strategists at time $t$, and let $y(t)$ represent the frequency of $T$-strategists. Obviously, the vector $(x(t), y(t))$ is a sufficient description of the state of the system. Given any such state, it is straightforward to compute the expected pay-offs obtained by each of the three strategies. Relying on obvious notation, they may be written as follows:

$$
\begin{aligned}
\pi(C^*; (x, y)) &= 3(1 - x); \\
\pi(D^*; (x, y)) &= 4 - 3x - 2y; \\
\pi(T; (x, y)) &= 3 - \tfrac{7}{3}x.
\end{aligned}
$$

Correspondingly, the mean expected pay-off across all three strategies is given by:

$$
\bar{\pi}(x, y) = 3 - 2x - \tfrac{4}{3}xy.
$$

This leads to the following Replicator Dynamics:

$$
\begin{aligned}
\dot{x}(t) &= x(t)\left[1 - x(t) - 2y(t) + \tfrac{4}{3}x(t)\,y(t)\right]; \\
\dot{y}(t) &= y(t)\left[-\tfrac{1}{3}x(t) + \tfrac{4}{3}x(t)\,y(t)\right].
\end{aligned}
$$

The stationary points of this dynamics are as follows. On the one hand, we know that the homogeneous profiles: $(1, 0)$, $(0, 1)$, and $(0, 0)$, are always stationary in the Replicator Dynamics. In addition, all points in the set $H \equiv \{(0, y) : 1 > y > 0\}$ are also stationary, i.e. the set of profiles where all players either play $C^*$ or $T$ in any arbitrary (positive) proportion. Finally, there is also an isolated stationary point $(\tilde{x}, 1 - \tilde{x})$ where $\tilde{x}$ is the unique point which satisfies:

$$
\pi(D^*, (\tilde{x}, 1 - \tilde{x})) = \pi(T, (\tilde{x}, 1 - \tilde{x})).
$$

In our example, $\tilde{x} = 3/4$.

Let us turn now to the stability of these stationary points. Since the analysis is straightforward, it is only sketched. First, it is clear that $(1,0)$ is asymptotically stable (recall Definition 6). Specifically, consider the neighbourhood of this point given by $U = \{(x,y) \in [0,1]^2 : x > \frac{3}{4}\}$. If $x(0) \in U$, we have that, for all $t \geq 0$, $\dot{x}(t) > 0$ and (if $y(t) > 0$) $\dot{y}(t) < 0$, which implies both Liapunov stability and convergence towards $(1,0)$.

On the other hand, the set $H \cup \{(0,0),(0,1)\}$ can be partitioned into two subsets, $H_1$ and $H_2$, separated by the point $(0,\hat{y})$ that satisfies:

$$\pi(C^*,(0,\hat{y})) = \pi(T,(0,\hat{y})), \tag{3.34}$$

i.e. where strategies $C^*$ and $T$ obtain identical pay-offs. In our example, we have $\hat{y} = \frac{1}{2}$. The points in the set $H_1 \equiv \{(0,y) : 1 \geq y \geq \hat{y}\}$ all satisfy Liapunov stability but fail nevertheless to satisfy the second convergence requirement of asymptotic stability. To see this, consider any $(0,y') \in H_1$ (and some given neighbourhood $V$ of it). If $(x(0),y(0)) \in V$ and $x(0) > 0$, $y(0) \geq y'$, then the ensuing trajectory satisfies:

$$\lim_{t\to\infty} x(t) = 0; \ \lim_{t\to\infty} y(t) = y^* > y'.$$

Thus, convergence to $(0,y')$ does *not* follow, no matter how small the neighbourhood $V$ is. However, if this neighbourhood is chosen small enough, it can be ensured that $|y^* - y'|$ is arbitrarily small, i.e. it belongs to any pre-specified neighbourhood of $(0,y')$. The reason for this is quite clear. The limit frequency of $T$-strategists will increase over its initial value as a continuous function on the frequency of $D^*$-strategists in the initial state. It is precisely this (small) frequency of $D^*$-strategists which induces the (progressively smaller) pay-off wedge between the pay-offs of $T$- and $C$-strategists along the trajectory. By a similarly straightforward argument, it may be concluded that neither the point $(\tilde{x}, 1 - \tilde{x})$ nor those in the set $H_2 \equiv H \setminus H_1 = \{(0,y) : 0 \leq y < \hat{y}\}$ are asymptotically stable. Indeed, they are not even Liapunov stable.

One may go beyond the purely local analysis summarized so far to find (*via* simulations) that the typical *global* dynamics of the Replicator Dynamics is as illustrated in Figure 2.

Thus, except for a one-dimensional manifold along which the system converges to the point $(\tilde{x}, 1-\tilde{x})$, any other point in the interior of the state space belongs to the basin of convergence of either the point $(1,0)$ or the set $H_1$. In fact, such a manifold separates those two basins of attraction. These stark dynamics illustrate in a simple fashion the crucial role played by the strategy tit-for-tat in the consolidation of co-operative behaviour (recall our former discussion). In particular, it indicates that only if the eventual proportion of individuals playing strategy $T$ is large enough (in particular, above $\hat{y}$) may a co-operative configuration prevail in the long run. This reflects the simple (but important) fact that, if co-operation is to survive in the long run, enough "reactive" individuals must persist in the population in order to protect otherwise exploitable "flat" co-operators.

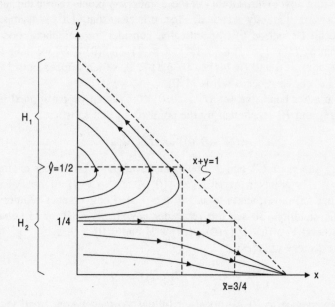

Figure 2: Repeated prisoner's dilemma, unperturbed dynamics

### 3.9.2    *Noisy dynamics*

The fact that the "co-operation component" $H_1$ merely specifies a lower bound on $y$ (thus allowing for a continuous range of different proportions between $C^*$ and $T$-strategists) implies that, close to this set, selection forces discriminating between strategies $C^*$ and $T$ must be very weak. In these circumstances, one should expect that, in the vicinity of the set $H_1$, small perturbations (or noise) may have a significant effect on the evolutionary dynamics.[19]

Noise may be introduced into the evolutionary dynamics in at least two different ways. One possibility is to contemplate aggregate-based noise that affects all individuals in the population in some correlated fashion. This would lead to the study of a genuinely stochastic dynamical system, as formulated in Section 5.5. Here, the alternative option is considered where noise is conceived as an individual-based phenomenon of "mutation". Assuming that it is statistically independent across individuals and time, we invoke the Law of Large Numbers to formulate a *deterministic* perturbation of the original Replicator Dynamics.

Specifically, we focus on a perturbation of the original system where, in

---

[19] Much of this subsection is heavily inspired by the work of Gale *et al.* (1995), applied to the so-called Ultimatum Game. Their approach is summarized in Section 4.8. The work by Hirshleifer and Martinez Coll (1992) for the repeated prisoner's dilemma also bears some relationship to the present discussion. Relying on numerical simulations, they reach similar conclusions.

a small but steady frequency, each one of the three strategies of the game is assumed to be introduced in the population irrespective of any pay-off (or selection) considerations. If $\theta > 0$ stands for the *mutation rate*, this amounts to transforming the original Replicator Dynamics as follows:

$$\dot{x}(t) = (1-\theta)\,x(t)\left[1 - x(t) - 2y(t) + \tfrac{4}{3}\,x(t)\,y(t)\right] \tag{3.35}$$
$$+\theta\left(\tfrac{1}{3} - x(t)\right);$$
$$\dot{y}(t) = (1-\theta)\,y(t)\left[-\tfrac{1}{3}\,x(t) + \tfrac{4}{3}\,x(t)\,y(t)\right] + \theta\left(\tfrac{1}{3} - y(t)\right). \tag{3.36}$$

The above expressions may be provided with the following interpretation. At every $t \geq 0$, each individual is subject to the same independent probability $(1-\theta)$ of "staying alive" and producing offspring in proportion to its respective pay-offs (i.e. in accordance with the Replicator Dynamics). With the complementary probability $\theta$, an individual dies (migrates or mutates), after which the individual that replaces it (perhaps itself, if it has only mutated) adopts some new strategy afresh, all three strategies selected with the same a priori probability of $\tfrac{1}{3}$.

As advanced, the introduction of such perturbation into the model has significant qualitative effects on its long-run performance. And this, it must be emphasized, occurs *even if* $\theta$ is chosen arbitrarily small. The most interesting effect takes place in a neighbourhood of the set $H_1$. Specifically, it will be shown below that, if $\theta$ is small, the analogue of this set now shrinks to just two points. Only one of them is locally stable, and converges to a point in $H_1$ as $\theta \to 0$. Thus, by introducing such mutational noise into the system, a clear-cut selection criterion is obtained which singles out a unique co-operative configuration in $H_1$.

To show this, first compute the stationary points of the perturbed Replicator Dynamics close to the set $H$ as follows. Make $\dot{x}(t) = \dot{y}(t) = 0$ in (3.35)–(3.36) and rewrite the resulting equations as follows:

$$(1-\theta)\,x\left[1 - x - 2y + \tfrac{4}{3}\,x\,y\right] = \theta\left(\tfrac{1}{3} - x\right); \tag{3.37}$$

$$(1-\theta)\,y\left[-\tfrac{1}{3}\,x + \tfrac{4}{3}x\,y\right] = \theta\left(\tfrac{1}{3} - y\right). \tag{3.38}$$

Dividing (3.37) by (3.38), and realizing that the value of $x$ that solves the above system of equations must be positive (since $\theta > 0$), we obtain the following equation:

$$\frac{1 - x - 2y + \tfrac{4}{3}xy}{\tfrac{4}{3}y^2 - \tfrac{1}{3}y} = \frac{x - \tfrac{1}{3}}{y - \tfrac{1}{3}}.$$

If $\theta$ is small (or, more formally, if we make $\theta \to 0$), the value of $x$ which solves the previous equation must converge to zero. Thus, making $x = 0$ in the above expression and rearranging terms, we have:

$$\frac{3 - 6y}{4y^2 - y} = \frac{1}{1 - 3y}. \tag{3.39}$$

Hence we can solve for the limit stationary values of $y$, close to $H$, which arise when $\theta$ becomes arbitrarily small. From (3.39), they can be seen to be the roots of the polynomial $y^2 - y + \frac{3}{14}$. That is:

$$y_1 = \tfrac{1}{2} + \tfrac{1}{2\sqrt{7}} \; ; \quad y_2 = \tfrac{1}{2} - \tfrac{1}{2\sqrt{7}} \; .$$

Notice that $y_1 > \hat{y} > y_2$, where recall that $\hat{y}\ (= 1/2)$ was defined by (3.34) as the minimum $T$-frequency in the set $H_1$. Thus, $y_1 \in H_1$ and $y_2 \in H_2$.

We finally argue that the stationary point $(x_1, y_1)$ is locally stable, whereas $(x_2, y_2)$ is not. Let $G : [0, 1]^2 \to \Re^2$ be the vector field which defines the perturbed dynamical system (3.35)–(3.36) and denote by $J_i \equiv DG(x_i, y_i)$ its Jacobian matrices evaluated at each of the two former stationary points $(x_i, y_i)$, $i = 1, 2$. We have:

$$J_i = \left( \begin{array}{cc} (1 - \theta)(1 - 2y_i - 2x_i + \frac{8}{3}x_i y_i) - \theta & (1 - \theta)(2x_i + \frac{4}{3}x_i^2) \\ (1 - \theta)(\frac{4}{3}y_i^2 - \frac{1}{3}y_i) & (1 - \theta)(\frac{8}{3}x_i y_i - \frac{1}{3}x_i) - \theta \end{array} \right) .$$

Since, for small $\theta$ and $x_i$, $1 - 2y_i$ dominates all other terms in the trace of each $J_i$ $(i = 1, 2)$, it follows immediately that the trace of $J_1$ is negative whereas that of $J_2$ is positive. This already discards $(x_2, y_2)$ as a potential asymptotically stable equilibrium of (3.35)–(3.36).[20] To confirm that $(x_1, y_1)$ is indeed asymptotically stable, it still needs to be checked that $|J_1| > 0$. By relying on the equilibrium conditions (3.37) and (3.38), it is easy to confirm that, for small $\theta$ (which allows us to ignore terms which are of second order in $\theta$ and/or $x_1$), the product of the main-diagonal elements of $J_1$ is positive, whereas that of its other two entries is negative.

This provides us with a clear picture of the *local* dynamics of the perturbed model close to the set $H$. The global picture elsewhere is not much affected (in comparison with that discussed in Subsection 3.9.1) by the introduction of small noise into the system. Figure 3 illustrates such global behaviour.

To summarize, we have shown that if the original Replicator Dynamics is perturbed by small mutational noise as described, only an *exact* frequency close to $y_1\ (> \frac{1}{2})$ of $T$-strategists will support a stable (in fact, an asymptotically stable) co-operative outcome. This contrasts with the wider range of less solid (i.e. only Liapunov stable) predictions arising from the unperturbed Replicator Dynamics. In a sense, what mutation does is to impose a test of robustness on these Liapunov stable equilibria which is able to select uniquely among them.

Nothing perhaps very new or surprising is added by this analysis to the insights already derived from the original model (especially, on the role played by $T$-strategists in the consolidation of co-operation). However, as mentioned, it illustrates the following important point, which will be repeatedly encountered in much of what follows (especially in Chapters 5 and 6): the (natural) introduction of noise into an evolutionary model often brings "order rather than chaos" into the model's behaviour.

---

[20] A two-dimensional square matrix $A$ has eigenvalues with negative real part if, and only if, its trace is negative and its determinant positive.

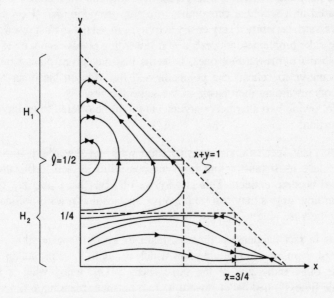

Figure 3: Repeated prisoner's dilemma, noisy dynamics

## 3.10 Pollination and Reward: An Example

### *3.10.1 Preliminaries*

As explained in Section 3.2, the basic interpretation of the Replicator Dynamics is of a biological nature, i.e. as a formalization of a Darwinian process of natural selection which reflects fitness – or reproductive – differentials.[21] Thus, before we turn to the study of *social* evolutionary systems in the ensuing chapters, it will be instructive to close this chapter with an application whose concern is strictly biological. Specifically, we shall focus on the study of pollination, which is a biological phenomenon which displays a strong "strategic flavour."

Pollination, as everyone knows, involves a balanced *quid pro quo* between plants and animals. Plants provide some substance which the animals value (say, nectar), thus inducing the latter to visit the former. This, in turn, allows plants to *cross*-fertilize each other by having the pollen attached to their visitors land in other members of their same species.

This is a rich phenomenon which can be, and has been, analysed from a wide variety of perspectives.[22] Here, we shall focus on the plants' perspective,

---

[21] See, however, Subsection 4.3.3 for a social interpretation (based on learning) of the Replicator Dynamics.

[22] See e.g. Selten and Shmida (1991) or Peleg and Shmida (1992).

taking the pollinators' behaviour as given. The question we shall ask is a simple one, phrased in a way that economists will find very familiar: How is it that offering reward, certainly a very costly activity, can persist in the presence of the acute free-rider problem it induces? If non-rewarding plants cannot be externally distinguished from rewarding ones, it seems that any given plant would have strong incentive to "cheat" the pollinator and free-ride on the other (say, the majority of) rewarding individuals of the same species.

The following two alternative arguments might be offered to tackle the previous question:

(a) If the plant species under consideration were not rewarding, it could not reproduce itself (abstracting from the possibility of self-fertilization) and would become extinct. This is why we observe that plants are typically rewarding, unless they can mimic other species that reward pollinators (or are otherwise able to deceive them).

(b) If (as in fact happens) pollinators learn to avoid a species that does not reward them, any individual plant which wants to have its pollen fertilize some other individual of the same species (i.e. enjoy what is labelled "male fitness") had better reward it. In a nutshell, plant male fitness is the incentive for reward.

The first tentative answer is, of course, utterly fallacious if interpreted to mean that what is "good for the species" must eventually evolve. It is precisely the desire to escape such teleological considerations which, to a great extent, has promoted the widespread use of Evolutionary Game Theory in modern biological analysis.

There is, however, an alternative interpretation for (a) which is based upon a hierarchical (or multi-level) view of the evolutionary process. That is, the view that evolution proceeding within each species at the "lower level" should be viewed as complemented by selection taking place across different species at some upper level. In the present case, however, this approach is highly problematic. It implicitly builds upon the untenable assumption that selection at the upper (interspecies) level is sufficiently fast to be relevant for the developments unfolding at the lower (intraspecies) level. As illustrated below (see Section 4.9), a hierarchic approach to evolution is much more natural in those contexts (mostly social or economic) where selection at the higher level operates on relatively small groups (e.g. firms), whose rate of selection (e.g. turnover) may be appropriately conceived as being relatively fast.

Let us now turn to (b) above. It focuses on what must surely be one important aspect of the phenomenon at hand, i.e. the incentives enjoyed by pollen-providing plants (in their male roles), aiming to have their pollen fertilize other flowers (in their female roles). The emphasis suggested by (b) is certainly important: pollinators must be provided with the right inducements to keep visiting plants of the same species. However, there is another dimension of the problem which (b) ignores that may be summarized as follows. For a rewarding plant, it is crucial to "assess" the likelihood that a rightly induced pollinator subsequently finds some other plant (i.e. flower) of the species which has *not yet* been fertilized. Of course, this likelihood must depend on how frequently plants are visited, and how often flowers are renewed during the season. In fact, one would expect that any analysis of the phenomenon (either theoretical or empirical) should display a positive relationship between flower turnover and the extent of rewarding behaviour. (See Primack (1985), for empirical evidence confirming this relationship.) A simple model with these characteristics (essentially taken from Vega-Redondo (1996*b*)) is subsequently proposed.

### 3.10.2    The model

Consider some given plant and pollinator species coexisting throughout every pollination season. These seasons are segmented into $T$ discrete periods, indexed by $t = 1, 2, ...T$. Every pollinator lives one whole season, its behaviour throughout it modelled exogenously according to the following simple behavioural rule:

> At any $t = 1, 2, ...T - 1$, if the visited plant has offered no reward, avoid the species during the following period. Otherwise, visit the species (once) again at $t + 1$.[23]

Plants (or flowers), on the other hand, may have a life-span $m$ ($1 \leq m \leq T$) shorter than the duration of the season. Specifically, we postulate an overlapping-generation structure of the following type. At the beginning of each $t = 1, 2, ..., T - m + 1$, there are $\xi$ flowers newly blooming which live for $m$ full periods. Thus, throughout the season, there is a total of $\xi(T - m + 1)$ flowers, their numbers at any given period ranging from a minimum of $\xi$ at the beginning and end of the season, to a maximum of $\xi m$ flowers at intermediate periods.

Plants can be of two types: rewarding (the $\alpha$-type) or non-rewarding (the $\beta$-type). Their respective frequencies will be denoted by $\nu_\alpha$ and $\nu_\beta$, respectively ($\nu_\alpha + \nu_\beta = 1$). The following three key assumptions will be made.

The first assumption specifies that reward is costly. That is, an $\alpha$-plant incurs a cost by rewarding that is reflected by a lower number of seeds produced when fertilized by incoming pollen (i.e. in its female role). Naturally, this cost is assumed to increase monotonically with the flower life-span. Formally,

---

[23] Implicit in this formulation is the idea that several plant species coexist throughout the season, any pollinator which is avoiding any given one of them visiting instead the other species. It is also convenient to make the assumption that the period length is so short that no pollinator can make more than one visit in each period.

*Assumption 1:* When fertilized, the $\beta$-plants produce $k$ times more seeds than the $\alpha$-plants, with $k = 1 + \psi(m)$, where $\psi(\cdot)$ is positive and non-decreasing.

The second assumption implicitly captures, in a very schematic fashion, what could be described as a pollinator-abundant scenario.[24] The underlying idea is that, when there are many pollinators relative to the number of plants, essentially all blooming flowers must be fertilized as long as *some* pollinators are being rewarded (and thus subsequently revisit the species). That is:

*Assumption 2:* If $\nu_\alpha > 0$, all flowers newly bloomed at $t = 2, 3, ..., T - m + 1$ are fertilized (i.e. produce seeds).

Finally, the third assumption concerns the "durability" of pollen.

*Assumption 3:* The pollen gathered at any $t = 1, 2, ..., T - 1$ is only suitable for fertilization at $t + 1$.

The previous assumption may be interpreted as reflecting, for example, a situation where the pollen is very weakly attached to the pollinator's body. In our analysis, it is connected to the assumed forgetfulness of at least one period by any "frustrated" pollinator. If pollinators were assumed to remember avoiding a non-rewarding species for more than one period, Assumption 3 could be relaxed accordingly.

We shall be concerned with the dynamics across seasons induced by the different rates of seed production enjoyed by each type of plant. Let seasons be indexed by $s = 1, 2, ...,$ and denote by $\nu(s) \equiv (\nu_\alpha(s), \nu_\beta(s))$ the profile of rewarding and non-rewarding plants prevailing in any given season. From Assumptions 1–3, the amounts of seeds produced of each type, $\alpha$ and $\beta$, must be proportional to the following respective magnitudes

$$n_\alpha(s) = \nu_\alpha(s) + \tfrac{1}{2}\nu_\beta(s)(1 + \psi(m)), \tag{3.40}$$

$$n_\beta(s) = \tfrac{1}{2}\nu_\beta(s)(1 + \psi(m)), \tag{3.41}$$

where the above expressions implicitly assume that every $\alpha$-pollen fertilizing a $\beta$-plant produces equal numbers of $\alpha$- and $\beta$-seeds. (Note that, given our assumption on pollinator behaviour, only $\alpha$-pollen will be available for fertilization.)

Further suppose that the numbers of *viable* seeds produced at the end of the season (i.e. those seeds which give rise to an "adult" plant at $s + 1$) are proportional to $n_\alpha(s)$ and $n_\beta(s)$.[25] Then, the average pay-off (or fitness) earned

---

[24] The symmetric pollinator-scarce context is analysed as well in Vega-Redondo (1996b), the main considerations involved shown to be very different.

[25] Thus, both $\alpha$- and $\beta$-seeds are affected by the same exogenous forces determining the probability that any given seed succesfully germinates.

by each strategy may be written as follows:

$$\pi_h(\nu(s)) = \kappa \, \frac{n_h(s)}{\nu_h(s)}, \quad h = \alpha, \beta,$$

for some given $\kappa > 0$. Hence, the discrete-time Replicator Dynamics (across seasons) induced by the model (cf. (3.3)) becomes:

$$
\begin{aligned}
\Delta \nu_h(s) &= \nu_h(s) \frac{\pi_h(\nu(s)) - [\nu_\alpha(s)\,\pi_\alpha(\nu(s)) + \nu_\beta(s)\,\pi_\beta(\nu(s))]}{\nu_\alpha(s)\,\pi_\alpha(\nu(s)) + \nu_\beta(s)\,\pi_\beta(\nu(s))} \\
&= \nu_h(s) \frac{n_h(s)/\nu(s) - [n_\alpha(s) + n_\beta(s)]}{n_\alpha(s) + n_\beta(s)}
\end{aligned}
$$

for each $h = \alpha, \beta$.

For the sake of simplicity, it will be assumed that the above system may be appropriately analysed (or approximated) by its continuous-time version:[26]

$$\dot{\nu}_h(s) = \nu_h(s) \frac{n_h(s)/\nu(s) - [n_\alpha(s) + n_\beta(s)]}{n_\alpha(s) + n_\beta(s)}, \quad h = \alpha, \beta, \tag{3.42}$$

for all $s \geq 0$.

The analysis will focus on issues of *global* stability. Specifically, a profile $(\nu_\alpha^*, \nu_\beta^*)$ will be said to be *globally evolutionarily stable (GES)* if for all *interior* initial conditions $(\nu_\alpha(0), \nu_\beta(0)) \in (0,1)^2$, any trajectory of the system $(\nu_\alpha(\cdot), \nu_\beta(\cdot))$ satisfies $(\nu_\alpha(\cdot), \nu_\beta(\cdot)) \to (\nu_\alpha^*, \nu_\beta^*)$. As shown below, our previous assumptions guarantee that the model always has a (unique) GES.

Let $\rho \equiv \frac{1}{m}$ denote the flower turnover rate (i.e. the fraction of flowers being renewed in every "central" period $t$ with $T - m + 1 \geq t \geq m + 1$). The main objective is to establish a clear-cut relationship between this turnover rate $\rho$ and the extent of rewarding behaviour (i.e. the frequency $\nu_\alpha^*$) prevailing at the GES state of the system. Such a relationship is contained in the following result.

**Theorem 11 (Vega-Redondo (1996b))** *Consider Assumptions 1–3, and let $\psi(1) \leq 1 < \psi(T)$. There exists some $\hat{\rho} \in (0,1)$ such that:*
*(a) if $\rho \geq \hat{\rho}$, the (unique) GES state is $(1,0)$;*
*(b) if $\rho < \hat{\rho}$, the (unique) GES state $(\nu_\alpha^*, \nu_\beta^*)$ satisfies $1 > \nu_\alpha^* > \nu_\beta^* > 0$.*

**Proof.** Define:

$$\phi(\nu_\alpha) \equiv \frac{n_\alpha/\nu_\alpha - n_\beta/(1 - \nu_\alpha)}{n_\alpha + n_\beta},$$

which, from (3.40) and (3.41), can be rewritten as

$$\phi(\nu_\alpha) = \frac{1}{1 + \nu_\beta(s)\psi(m)} \left[ 1 + \frac{1}{2}(1 + \psi(m))(\frac{\nu_\beta(s)}{\nu_\alpha(s)} - 1) \right].$$

---

[26] As explained in Subsection 3.2.2, we may assume that the base fitness of both types is very high, implying that their relative fitness differentials within each season are very low.

Consider first the case where $\rho = m = 1$. Then, since $\psi(1) \leq 1$, we must have $\phi(\nu_\alpha) > 0$ for all $\nu_\alpha < 1$. Since, from (3.42),

$$\frac{\dot{\nu}_\alpha(s)}{\nu_\alpha(s)} - \frac{\dot{\nu}_\beta(s)}{\nu_\beta(s)} = \phi(\nu_\alpha(s)),$$

it follows that $[1,0]$ is the unique GES state for this case. Clearly, the same applies for all $\rho$ such that $\psi(\frac{1}{\rho}) \leq 1$. Choose $\hat{\rho}$ as the smallest $\rho$ for which the previous inequality applies. Since $\psi(T) > 1$, such $\hat{\rho} > 0$ exists, and the proof of part (a) is complete.

Consider now the case where $\rho < \hat{\rho}$. Then, $\phi(1) < 0$, which implies that $[1,0]$ is not a GES state. On the other hand, we have $\phi(\frac{1}{2}) > 0$ and, therefore, all interior stationary points of the dynamics must have a frequency of $\alpha$-plants larger than $\frac{1}{2}$. In fact, since $\phi(\cdot)$ is a monotonic function in $\nu_\alpha$ there is a unique $\nu_\alpha^* \in (\frac{1}{2}, 1)$ such that $\phi(\nu_\alpha^*) = 0$. Since, moreover, $\phi(\cdot)$ is strictly decreasing (cf. Figure 4), we conclude that $(\nu_\alpha^*, 1 - \nu_\alpha^*)$ is a GES state. This completes the proof. ∎

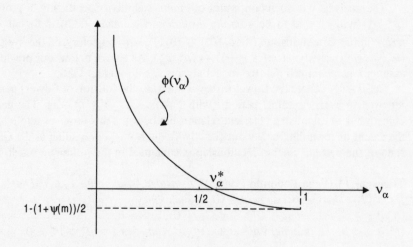

Figure 4: Differential-fitness function $\phi$

The previous result makes transparent the necessary balance between the two dimensions of plant fitness (male- and female-based) which must underlie any evolutionarily stable state in a pollination environment. It also clarifies the fact that, in a pollinator-abundant context, this balance has to depend crucially on the rate of flower turnover prevailing throughout the pollination season.

# 4

# Evolution in Social Environments

## 4.1  Introduction

The previous chapter has focused on the Replicator Dynamics (RD), a dynamic evolutionary system whose interpretation is of a strict biological nature. As explained in Section 3.2, it represents a direct formalization of a Darwinian process of selection, i.e. a process by which those strategies that prevail in the long run are merely those that reproduce faster.

Subsection 4.3.3 below illustrates that one may also attribute to the RD a certain "social interpretation" as the outcome of very specific processes of imitation or other particular forms of bounded-rationality decision making. In general, however, the RD formalization is too rigid to encompass the wealth of social evolutionary processes that will interest us. In most social and economic contexts, the underlying evolutionary process can only be determined at a broad qualitative level. That is, it is only possible to specify certain general features of the process, without being able to pinpoint any precise formulation for it. For example, we may be able to assert that actions which enjoy a relatively higher pay-off tend to spread, without being in a position to formulate any precise or cardinal measure of the extent to which this happens. In a natural sense, processes which display such qualitative behaviour should still be labelled evolutionary. However, a rigorous analysis of them requires the formulation of a theoretical framework which admits a much larger flexibility than that studied in the previous chapter.

This will be one of the concerns and motivations for the present chapter. A second one derives from our desire to tackle contexts where, unlike what was postulated in the previous chapter, individuals *cannot* be suitably viewed as members of a single population. In many interesting situations, some members of the population (say, buyers) will not interact among themselves but rather do it with some other different set of individuals (i.e. sellers). In order to study these contexts, we need to consider a richer theoretical framework that potentially allows for several distinct populations, their respective members cross-interacting in some particular way.

The two issues outlined require a substantial generalization of the evolutionary approach studied in the preceding two chapters. The pay-off will also be substantial. We shall enlarge the range of application of the theory to many social and economic environments for which a generalized evolutionary approach will provide new and important insights.

## 4.2    Theoretical Framework

For simplicity, we shall just consider two populations, 1 and 2. (Most of the analysis may be extended to any finite number of populations, at the cost merely of notational complexity.) Time $t$ is measured continuously, $t \in [0, \infty)$. At each $t$, every member of each population is randomly matched with individuals from the other population in order to play a bilateral finite game with pay-off matrices $A, B \in \Re^{m_1 \times m_2}$. Here, $a_{ij} \equiv \pi^1 \left( s_i^1, s_j^2 \right)$ and $b_{ij} \equiv \pi^2 \left( s_i^1, s_j^2 \right)$ stand for the pay-offs obtained by a player of type 1 (population 1) and a player of type 2 (population 2), respectively, if the former adopts strategy $s_i^1$ and the latter strategy $s_j^2$.

The strategy spaces of each type are denoted by $S^1$ and $S^2$, with respective cardinalities $m_1$, $m_2 \in \mathcal{N}$. Correspondingly, $\Sigma^1 = \Delta^{m_1-1}$ and $\Sigma^2 = \Delta^{m_2-1}$ stand for their respective spaces of mixed strategies, with generic elements $\sigma^1 \in \Sigma^1$ and $\sigma^2 \in \Sigma^2$. The pay-off functions $\pi^1$ and $\pi^2$ are extended to these latter spaces in the usual fashion.

If individuals are assumed to adopt only pure strategies (which will be assumed for most of this chapter), the spaces $\Delta^{m_1-1}$ and $\Delta^{m_2-1}$ can also be interpreted as the set of possible population states (i.e. population profiles specifying the fraction of individuals playing each of the different pure strategies). Such population states will be generically denoted by $\nu^1$ and $\nu^2$, where $\nu_i^k$ specifies the fraction of individuals of population $k = 1, 2$ which adopt strategy $s_i^k \in S^k$. Under our assumption of random matching, we shall abuse previous notation and denote by $\pi^k \left( s_i^k, \nu^{k'} \right)$, $k, k' = 1, 2$, $k \neq k'$, the expected pay-off of playing strategy $s_i^k$ for an individual of population $k$ when the strategy profile prevailing in the other population $k'$ is $\nu^{k'}$. Further assuming that the cardinality of each population is the continuum, such expected values are also identified with the corresponding average magnitudes.

## 4.3 Evolutionary Growth Dynamics

### 4.3.1 The model

For the moment, the analysis will focus on dynamical systems on $\Omega \equiv \Delta^{m_1-1} \times \Delta^{m_2-1}$ of the form:

$$\dot{\nu}_i^k(t) = \nu_i^k(t) \, F_i^k \left( \nu^1(t), \nu^2(t) \right); \; i = 1, 2, ..., m_k; \; k = 1, 2, \qquad (4.1)$$

where every $F_i^k : \Omega \to R$ is assumed Lipschitz-continuous.[1] To be well defined, any such evolutionary system has to leave the state space $\Omega$ invariant, i.e. every trajectory induced by (4.1) must satisfy:

$$\nu(0) \equiv \left( \nu^1(0), \nu^2(0) \right) \in \Omega \Rightarrow \forall t \geq 0, \; \nu(t) \in \Omega.$$

To guarantee such invariance, it is enough to assume that, for each $k = 1, 2$,

$$\nu^k \cdot F^k(\nu) = \sum_{i=1}^{m_k} \nu_i^k F_i^k(\nu) \equiv 0,$$

i.e. all vectors of change induced by (4.1) belong to the tangent space of $\Omega$.

When the frequency of a certain strategy $\nu_i^k(t)$ is positive, the respective $F_i^k \left( \nu^1(t), \nu^2(t) \right)$ determines its current rate of change. This focus on rates of change may indeed be a suitable one for many social applications (see Subsection 4.3.3 for some examples). In other cases, however, it imposes undesirable restrictions. For example, it forces every strategy which is extinct at the beginning of the process to remain so throughout irrespective of any pay-off considerations. (Formally, this is a consequence of the fact that, as for the RD, the dynamical system (4.1) renders every face of the two simplices in $\Omega$ invariant.) This issue will be addressed in Section 4.6 by postulating a more general approach which is not subject to this limitation.

### 4.3.2 Monotonicity properties

Naturally, in order to think of any dynamical system as "evolutionary", it must be required to reflect some appropriate idea of *selection*. In the literature, this usually goes under the label of "monotonicity", with different qualifiers to this term indicating alternative variations of the general idea. A quite weak monotonicity concept (see Nachbar (1990)) is provided by the following definition.

---

[1] A function $f : \Re^n \to \Re$ is said to be Lipschitz-continuous if $\exists k > 0$ such that $\forall x, y \in \Re^n$, $|f(x) - f(y)| \leq k \, |x - y|$, where $|\cdot|$ denotes the Euclidean norm. By a standard result in the Theory of Differential Equations, this condition is sufficient to guarantee the existence and uniqueness of a solution to the dynamical system.

**Definition 10** *An evolutionary system (4.1) is said to be growth-monotonic if*
$\forall k, k' = 1, 2 \ (k \neq k'), \forall i, j = 1, 2, ..., m_k, \forall \nu = (\nu^1, \nu^2) \in \Omega,$[2]

$$\pi^k \left( s_i^k, \nu^{k'} \right) \geq \pi^k \left( s_j^k, \nu^{k'} \right) \Leftrightarrow F_i^k(\nu) \geq F_j^k(\nu).$$

Verbally, an evolutionary system is called *growth-monotonic* if the differential rates of change between any two pairs of strategies $s_i^k$ and $s_j^k$ which are
played in positive frequency:

$$\frac{\dot{\nu}_i^k(t)}{\nu_i^k(t)} - \frac{\dot{\nu}_j^k(t)}{\nu_j^k(t)} = F_i^k(\nu) - F_j^k(\nu)$$

is of the same sign as their differential pay-offs $\pi^k \left( s_i^k, \nu^{k'} \right) - \pi^k \left( s_j^k, \nu^{k'} \right)$.
This concept captures a very flexible notion of selection by which any strategy $s_i^k$
which does relatively better (worse, or equal) than any other $s_j^k$ within population
$k$ is simply required to have its relative frequency $\nu_i/\nu_j$ grow (respectively, fall
or remain equal), unless prevented from doing so by boundary considerations.

Growth monotonicity focuses on *dichotomic* comparisons between every pair
of strategies. It generalizes the idea captured by the quotient representation of
RD which, as explained in Subsection 3.2.3, is fully equivalent to the original
formulation of the RD.

In the original specification of the RD (3.4), the rate at which the share
of each strategy grows is linked to the difference between its current pay-off
and the average pay-off. The next definition presents an alternative monotonicity concept proposed by Ritzberger and Weibull (1995) which generalizes this
alternative formulation of the RD.

**Definition 11** *An evolutionary system (4.1) is said to be sign-preserving if*
$\forall k, k' = 1, 2 \ (k \neq k'), \forall i = 1, 2, ..., m_k, \forall \nu = (\nu^1, \nu^2) \in \Omega,$

$$\left[ \pi^k \left( s_i^k, \nu^{k'} \right) - \sum_{j=1}^{m_k} \nu_j^k \, \pi^k \left( s_j^k, \nu^{k'} \right) \geq 0 \right] \Leftrightarrow F_i^k(\nu) \geq 0.$$

Thus, an evolutionary system is *sign-preserving* if for any given pure strategy to increase its frequency it is both necessary and sufficient that its pay-off
be above the average one for the respective population.[3] Clearly, there is no

---

[2] Here, we abuse slightly previous notation since, when writing $\pi^k(s_i^k, \nu^{k'})$, the first argument
always corresponds to population $k$ (even if $k = 2$).

[3] Note that the statement of equivalence included in Definition 11 can be reformulated in terms
of the *reverse* strict inequalities. In fact, the same applies to the equivalence included in Definition
10). In this latter case, however, one need not reverse the strict inequalities due to to the symmetry
of the expression across different pairs of indices.

relationship of logical inclusion between the monotonicity conditions of Definitions 10 and 11. Both of them, of course, are satisfied by the "benchmark model" provided by the two-population RD:

$$\dot{\nu}_i^k(t) = \nu_i^k(t) \left[ \pi^k \left( s_i^k, \nu^{k'}(t) \right) - \sum_{j=1}^{m_k} \nu_j^k(t) \, \pi^k \left( s_j^k, \nu^{k'}(t) \right) \right], \qquad (4.2)$$

for each $i = 1, 2, ..., m_k$ where $k, k' = 1, 2 \; (k \neq k')$.

### 4.3.3    Some examples

We now discuss two alternative classes of examples which respond to very different underlying motivations. Within both of them, however, a particular (quite special) subclass induces a Replicator Dynamics which has a social-based (rather than biological) underpinning.

#### 4.3.3.1    Imitation dynamics

Consider two given populations with a continuum of individuals, matched every period to play some bimatrix game as described above. Suppose that, at every $t \geq 0$, an "infinitesimal" fraction of individuals in each population have the option of revising their strategy. In doing so, they adopt the following two-step procedure.

First, each one of them meets at random somebody else of their *own* population and observes both the latter's action, say $s'$, and associated pay-off, say $\pi'$. Second, each one compares his own action and pay-off, $s$ and $\pi$, with the realized observation. If it happens that $s' \neq s$ and $\pi' > \pi$, the individual in question is assumed to shift from $s$ to $s'$ with some probability $p(s, s', \pi', \pi) > 0$ which, in general, is some function of the strategies and pay-offs involved. Otherwise (i.e. if $s = s'$ or $\pi' \leq \pi$), he is assumed to continue playing the original strategy $s$. Here, strategies could be identified with technologies, degrees of altruism, or languages, all of them adopted to some extent by imitation in social environments.

Assume, for simplicity, that the number of random matchings per period is large so that each individual's average pay-off may be well approximated by the corresponding expected magnitude. Then, if $p(\cdot)$ is symmetric across strategies and non-decreasing in payoff gaps, the resulting adjustment process yields a growth-monotonic evolutionary system when formulated in continuous time. For, in this case, strategies which yield a higher average pay-off are freshly adopted by more individuals (and abandoned by fewer) than those with lower average pay-off.

The following interesting interpretation for the revision probability $p(\cdot)$ has been suggested by Nachbar (1990). Assume that individuals evaluate their strategies solely in terms of short-run pay-offs. Then, as long as $\pi' > \pi$ (with the above interpretation), it will pay for an individual to shift to strategy $s'$ with *probability one* if there are no adjustment costs and the population adjustment is very gradual (thus, current pay-offs are good predictors of those following next). Suppose, however, that individuals are subject to a switching cost $c$, which has to be instantaneously paid if an individual changes his strategy. Furthermore, suppose that this switching cost is not fixed and deterministically given but, rather, is independently determined across individuals and time according to some common probability measure on $\Re_+$. Let $D(\cdot)$ be its cumulative distribution function. Then, any individual whose observation of a strategy $s' (\neq s)$ is associated with a higher pay-off $\pi'$ will adopt $s'$ only if $c \leq \pi' - \pi$. This will occur with a probability:

$$p(s, s', \pi', \pi) = D(\pi' - \pi).$$

Interestingly enough, if the framework described is now specialized even further by assuming that the probability distribution over switching costs is uniform with support on a sufficiently large interval $[0, M]$, the resulting evolutionary system coincides with the Replicator Dynamics. Specifically, it is enough that $M \geq 2 \max\{|a_{ij}|, |b_{ij}| : i = 1, 2, ..., m_1, \ j = 1, 2, ..., m_2\}$. In this case, $D(\pi - \pi') = (\pi - \pi')/M$ and the contemplated formulation induces the following population dynamics:

$$\dot{\nu}_i^k(t) = -\nu_i^k(t) \sum_{j=1}^{m_k} \nu_j^k(t)(1/M) \left[\pi^k\left(s_j^k, \nu^{k'}(t)\right) - \pi^k\left(s_i^k, \nu^{k'}(t)\right)\right]^+$$
$$+ \sum_{j=1}^{m_k} \nu_j^k(t) \, \nu_i^k(t) \, (1/M) \left[\pi^k\left(s_i^k, \nu^{k'}(t)\right) - \pi^k\left(s_j^k, \nu^{k'}(t)\right)\right]^+,$$

$$(4.3)$$

where $[x]^+ \equiv \max\{x, 0\}$. The first term in (4.3) represents the rate at which $s_i^k$-adopters, on average, quit using this strategy. On the other hand, its second term reflects the average rate at which the individuals of population $k$ switch to (or remain with) strategy $s_i^k$. Immediate algebraic manipulation of (4.3) yields:

$$\dot{\nu}_i^k(t) = (1/M) \, \nu_i^k(t) \left[\pi^k\left(s_i^k, \nu^{k'}(t)\right) - \sum_{j=1}^{m_k} \nu_j^k(t) \, \pi^k\left(s_j^k, \nu^{k'}(t)\right)\right]$$

for each $i = 1, 2, ..., m_k$, $k, k' = 1, 2 \ (k \neq k')$. Modulo some "scale factor" $1/M$ (which, being *common* to both populations, does not affect the qualitative behaviour of the system),[4] it coincides with the two-population RD (4.2).

---

[4] Recall our discussion in Subsection 3.2.2.

### 4.3.3.2 Satisficing dynamics

Consider now a different approach which is a slight variation of a formulation proposed by Cabrales (1992), itself inspired in a model of Smallwood and Conlisk (1979). In contrast with the preceding context, individuals are assumed to learn nothing at all about the pay-offs obtained by others. They only observe the average pay-off $\pi$ which they earn from their current strategy $s$, and compare it to some target level of satisfaction $\mu$. If $\pi \geq \mu$, they stay with strategy $s$. Otherwise, they choose some new strategy, the probability with which any given $s_i$ is then selected being equal to the frequency of individuals $\nu_i$ who adopt it within the corresponding population.

In the context just outlined, the probability that a certain strategy is selected *afresh* only depends on the relative frequencies with which it is currently adopted in the corresponding population, *not* on its average pay-off. Pay-offs, in this case, are only relevant in triggering the decision of changing strategies when the target level of satisfaction is not met.

To complete the model, we need to specify how individuals determine their satisfaction target level. Let us assume that it is determined stochastically according to some common probability measure on $\Re$, independent across individuals and time, and with cumulative distribution function $Y(\cdot)$. One possible interpretation here is that the "mood" of an individual (e.g. whether he feels ambitious or not) is determined stochastically, depending on a set of variable idiosyncratic conditions. In any case, it should be clear that, under gradual adjustment, i.e. when only a "very small" (infinitesimal) fraction of individuals may revise their strategy at each point in (continuous) time, the postulated framework yields a growth-monotonic evolutionary system.

As before, let us now assume that $Y(\cdot)$ defines a uniform probability measure on a sufficiently large interval $[\alpha, \beta] \subset \Re$. Specifically, suppose that

$$\beta \geq \max\{a_{ij}, b_{ij} : i = 1, 2, ..., m_1, \ j = 1, 2, ..., m_2\}$$
$$\alpha \leq \min\{a_{ij}, b_{ij} : i = 1, 2, ..., m_1, \ j = 1, 2, ..., m_2\}.$$

Given any prevailing state $\nu$, consider an individual of population $k = 1, 2$ who is in a position to revise his current strategy $s_i^k \in S^k$ and whose current pay-off is $\pi$. With probability $\frac{\beta - \pi}{\beta - \alpha}$ this individual is dissatisfied with his strategy and will therefore revise it. In this case, the probability of choosing any particular $s_j^k \in S^k$ coincides with $\nu_j^k$, the current frequency of $s_j^k$-adopters in population $k$. Averaging over the individuals of population $k$ who are currently adopting strategy $s_i^k$ and have the option of revising their strategy, the following law of motion results:

$$
\dot{\nu}_i^k(t) = -\nu_i^k(t) \frac{\beta - \pi^k\left(s_i^k, \nu^{k'}(t)\right)}{\beta - \alpha}
$$
$$
+ \nu_i^k(t) \sum_{j=1}^{m_k} \nu_j^k(t) \frac{\beta - \pi^k\left(s_j^k, \nu^{k'}(t)\right)}{\beta - \alpha},
$$

$$(4.4)$$

where, as before, the first term of (4.4) represents the rate at which $s_i^k$-adopters choose a new strategy, whereas its second term captures those who adopt $s_i^k$ when revising their prior strategy. Straightforward manipulation of (4.4) yields:

$$\dot{\nu}_i^k(t) = \frac{1}{\beta - \alpha}\, \nu_i^k(t) \left[ \pi^k\left( s_i^k, \nu^{k'}(t)\right) - \sum_{j=1}^{m_k} \nu_j^k(t)\, \pi^k\left( s_j^k, \nu^{k'}(t)\right) \right],$$

for each $i = 1, 2, ..., m_k$, $k, k' = 1, 2$ $(k \neq k')$, again a mere scaling of the two-population RD.

## 4.4    Dynamics of Monotonic Evolutionary Systems

### 4.4.1    *Dynamic stability and Nash equilibrium*

The stability analysis of growth-monotonic (GM) or sign-preserving (SP) evolutionary systems with several populations displays some parallels, but also some essential differences from that conducted in Chapter 3 for a single population.

On the one hand, as before (cf. Proposition 7), asymptotic stability is a sufficient condition for Nash equilibrium, as stated by the following result.[5]

**Proposition 9 (Nachbar (1990))** *Let $\hat{\nu} \equiv \left(\hat{\nu}^1, \hat{\nu}^2\right)$ be an asymptotically stable point of a GM evolutionary dynamics (4.1). Then, as an element of $\Sigma = \Sigma^1 \times \Sigma^2$, $\hat{\nu}$ is a Nash equilibrium of the bimatrix game $(A, B)$.*

**Proof.** The proof is completely analogous to that of Proposition 7. Consequently, it is left to the reader.    ■

**Remark 5** *It should be apparent that the same conclusion of Proposition 9 applies to any asymptotically stable point of an SP evolutionary system.*

A more interesting question is polar to that addressed above, i.e. under what conditions is a Nash equilibrium asymptotically stable? In general, as we shall show below, only the very restrictive class of strict Nash equilibria may be asymptotically stable. To establish this negative claim, it is enough to show that it applies to some interesting set of cases. For the sake of concreteness, we shall focus on the most paradigmatic representative of the class of monotonic evolutionary systems: the Replicator Dynamics, as given by (4.2).

First, it is shown that only pure-strategy Nash equilibria can be asymptotically stable.

---

[5] In analogy with Proposition 6, it is clear that the set of Nash equilibria induces a subset (in general a proper one) of the set of rest points of a growth-monotonic evolutionary system.

**Theorem 12 (Eshel and Akin (1983))** *Let $\nu^* \in \Delta^{m_1-1} \times \Delta^{m_2-1}$ be an asymptotically stable equilibrium of the RD. Then, each $\nu^{k*}$, $k = 1, 2$, is a vertex of its respective $\Delta^{m_k-1}$.*

**Proof.** Let $B^k \equiv \{\nu^k \in \Re_+^{m_k-1} : \sum_{i=1}^{m_k-1} \nu_i^k \leq 1\}$, $B \equiv B^1 \times B^2$, and conceive the RD as a $(m_1 + m_2 - 2)$-dimensional vector field on $B$,

$$H \equiv H^1 \times H^2 : B \to \Re^{m_1+m_2-2}. \tag{4.5}$$

Assume, for the sake of contradiction, that there is an interior equilibrium which is asymptotically stable and let $A$ be an open neighbourhood of it. Based on $H(\cdot)$, define the instrumental vector field $Q : \text{int}(B) \to \Re^{m_1+m_2-2}$ as follows:

$$Q(\nu) = \frac{1}{P(\nu)} H(\nu),$$

where:

$$P(\nu) \equiv \left(\prod_{i=1}^{m_1-1} \nu_i^1\right) \left(1 - \sum_{i=1}^{m_1-1} \nu_i^1\right) \left(\prod_{i=1}^{m_2-1} \nu_i^2\right) \left(1 - \sum_{i=1}^{m_2-1} \nu_i^2\right).$$

Obviously, $Q(\cdot)$ and $H(\cdot)$ have the same qualitative dynamic behaviour (in particular, the same stable equilibria).

Now, we apply Liouville's Theorem (see the Appendix) on the vector field $Q(\cdot)$ to assert that the volume $V(t)$ of the set $A(t) = \{y = \hat{\nu}(t) : \hat{\nu}(0) \in A\}$ satisfies:

$$\dot{V}(t) = \int_{A(t)} \text{div } Q(\nu) \, d\nu, \tag{4.6}$$

where:

$$\text{div } Q(\nu) = \sum_{i=1}^{m_1-1} \frac{\partial Q_i^1(\nu)}{\partial \nu_i^1} + \sum_{i=1}^{m_2-1} \frac{\partial Q_i^2(\nu)}{\partial \nu_i^2} \tag{4.7}$$

is the divergence of the vector field $Q$ at $\nu$. We compute:

$$\frac{\partial Q_i^k(\nu)}{\partial \nu_i^k} = \frac{1}{P(\nu)} \frac{\partial H_i^k(\nu)}{\partial \nu_i^k} - \frac{1}{(P(\nu))^2} \frac{\partial P(\nu)}{\partial \nu_i^k} H_i^k(\nu);$$

$$\frac{\partial P(\nu)}{\partial \nu_i^k} = \frac{1 - \sum_{j=1}^{m_k-1} \nu_j^k - \nu_i^k}{\nu_i^k \left(1 - \sum_{j=1}^{m_k-1} \nu_j^k\right)} P(\nu).$$

Therefore,

$$\text{div } Q(\nu) = \sum_{k=1,2} \sum_{i=1}^{m_k-1} \frac{1}{P(\nu)} \left\{ \frac{\partial H_i^k(\nu)}{\partial \nu_i^k} - \frac{1 - \sum_{j=1}^{m_k-1} \nu_j^k - \nu_i^k}{\nu_i^k \left(1 - \sum_{j=1}^{m_k-1} \nu_j^k\right)} H_i^k(\nu) \right\}$$

$$= \frac{1}{P(\nu)} \left\{ \sum_{k=1,2} \left[ \begin{array}{c} div\, H(\nu) - \\ \sum_{i=1}^{m_k-1} \frac{H_i^k(\nu)}{\nu_i^k} - \sum_{i=1}^{m_k-1} \frac{H_i^k(\nu)}{\left(1 - \sum_{j=1}^{m_k-1} \nu_j^k\right)} \end{array} \right] \right\}.$$

Since, for each $k = 1, 2$, $i = 1, 2, ..., m_k - 1$,

$$H_i^k(\nu) = \nu_i^k \left( \pi^k \left( s_i^k, \nu^{k'} \right) - \pi^k \left( \nu^k, \nu^{k'} \right) \right)$$

and, for each $k = 1, 2$,

$$\sum_{i=1}^{m_k-1} \frac{H_i^k(\nu)}{\left(1 - \sum_{j=1}^{m_k-1} \nu_j^k\right)} = \frac{1}{\nu_{m_k}^k} \left\{ \begin{array}{c} \left[ \sum_{i=1}^{m_k-1} \nu_i^k \pi^k \left( s_i^k, \nu^{k'} \right) \right] \\ -(1 - \nu_{m_k}^k) \pi^k \left( \nu^k, \nu^{k'} \right) \end{array} \right\}$$

$$= \frac{1}{\nu_{m_k}^k} \left\{ \begin{array}{c} \left[ \pi^k \left( \nu^k, \nu^{k'} \right) - \nu_{m_k}^k \pi^k \left( s_{m_k}^k, \nu^{k'} \right) \right] \\ -(1 - \nu_{m_k}^k) \pi^k \left( \nu^k, \nu^{k'} \right) \end{array} \right\}$$

$$= -\left( \pi^k \left( s_{m_k}^k, \nu^{k'} \right) - \pi^k \left( \nu^k, \nu^{k'} \right) \right)$$

we have

$$div\, Q(\nu) = \frac{1}{P(\nu)} \left\{ div\, H(\nu) - \sum_{k=1,2} \sum_{i=1}^{m_k} \left( \pi^k \left( s_i^k, \nu^{k'} \right) - \pi^k \left( \nu^k, \nu^{k'} \right) \right) \right\}.$$

On the other hand, it is easy to calculate that:

$$div\, H(\nu) = \sum_{k=1,2} \sum_{i=1}^{m_k} \left( \pi^k \left( s_i^k, \nu^{k'} \right) - \pi^k \left( \nu^k, \nu^{k'} \right) \right).$$

Thus, we conclude that $div\, Q(\nu) = 0$ and, from (4.6) and (4.7), $V(t) = V(0) > 0$. This is obviously incompatible with asymptotic stability of the contemplated equilibrium. It follows, therefore, that no equilibrium interior to $\Omega$ (or $B$) can be asymptotically stable. But clearly, the same argument can be applied to any product of faces in the product simplex $\Omega \equiv \Delta^{m_1-1} \times \Delta^{m_2-1}$. Thus, any asymptotically stable equilibrium of $H(\cdot)$ must be a vertex of $\Omega$, i.e. must involve only pure strategies. ∎

From the previous result, the conclusion that every asymptotically stable state corresponds to a strict Nash equilibrium (cf. van Damme (1987)) follows as an easy corollary.

**Corollary 2 (Ritzberger and Vogelsberg (1990))** *Let $\nu^* \in \Delta^{m_1-1} \times \Delta^{m_2-1}$ be an asymptotically stable equilibrium of the RD. Then, as an element of $\Sigma = \Sigma^1 \times \Sigma^2$, $\nu^*$ is a strict Nash equilibrium of the bimatrix game $(A, B)$.*

**Proof.** Let $\left(s_i^1, s_j^2\right)$ correspond to an asymptotically stable pair of vertices and assume it does not define a strict Nash equilibrium. That is, there is for one of the populations, say population 1, another pure strategy $s_r^1$, $r \neq i$, such that $\pi^1\left(s_r^1, s_j^2\right) = \pi^1\left(s_i^1, s_j^2\right)$. Define $\delta_h^k$ to be the vector in $\Delta^{m_k-1}$ whose $h$th component equals one. Then, the following set of states

$$\chi = \{\nu \in \Omega : \nu^2 = \delta_j^2; \ \nu^1 = \lambda\delta_i^1 + (1-\lambda)\ \delta_r^1, \ \lambda \in [0,1]\}$$

are all rest points of $H\left(\cdot\right)$, as introduced in (4.5). This obviously contradicts the fact that $\left(s_i^1, s_j^2\right)$ is asymptotically stable, which completes the proof of the corollary. ∎

In contrast with the analysis undertaken in Chapter 3 (recall, for example, Theorem 3) the preceding results underscore the point that asymptotic stability is too restrictive a requirement in a context with several populations. In most interesting contexts, equilibria will typically involve some degree of polymorphism and, therefore, cannot be associated with a strict Nash equilibrium of the underlying game. In these circumstances, dynamic analysis cannot rely on the study of stable configurations. Instead, it may undertake two different (but complementary) approaches. On the one hand, it may extend the stability analysis to subsets of the state space rather than single points. As a second alternative, the analysis can aim at identifying certain long-run regularities displayed by dynamic trajectories, even if these are not convergent. Both approaches are illustrated in turn by the following two subsections.

### 4.4.2    Set stability

By suitably adapting Definition 6, the notion of asymptotic stability can be extended to any closed subset of the state space (not just points) in the obvious fashion. Relying on such a set-stability concept and focusing on sign-preserving evolutionary systems (SPES), Ritzberger and Weibull (1995) pose the following general question: Can sets which are asymptotically stable be usefully characterized, even if the particular evolutionary system in operation (only guaranteed to be an SPES) is not known ?

To tackle this issue, they restrict their attention to those sets which are Cartesian products of (whole) simplex faces. Let $X^k \subset S^k$ stand for some given subset of pure strategies for population $k$ and make

$$\Delta(X^k) \equiv \{\nu^k \in \Delta^{m_k-1} : s_i^k \notin X^k \Rightarrow \nu_i^k = 0\}.$$

Further denote $X \equiv X^1 \times X^2$ and $\Delta(X) \equiv \Delta(X^1) \times \Delta(X^2)$. An easy adaptation of the proof of Theorem 12 leads to the conclusion that, for the RD, every asymptotically stable set must be of the form $\Delta(X)$. Since the RD represents a special benchmark in evolutionary analysis (which is obviously sign-preserving – recall Subsection 4.3.2), restricting to sets of this form is a natural course to take in shedding some light on the previous question.

A key role in the present analysis is played by the so-called *better-reply correspondence* $\gamma = (\gamma^k)_{k=1,2} : \Omega \longrightarrow S^1 \times S^2$, defined as follows:

$$\gamma^k(\nu^1, \nu^2) = \{ s_i^k \in S^k : \pi^k(s_i^k, \nu^{k'}) \geq \pi^k(\nu^k, \nu^{k'}) \}.$$

In contrast with the standard best-response correspondence, $\gamma(\cdot)$ reflects the *relative* (rather than absolute) pay-off considerations which must underlie evolutionary analysis. In the spirit of the SPES concept, it relies on average pay-offs as the benchmark of comparison.

Heuristically, it seems clear that a necessary condition for any product simplex $\Delta(X)$ to be asymptotically stable with respect to a SPES must be that $\gamma(\Delta(X)) \subseteq X$, i.e. for any $(\nu^1, \nu^2) \in \Delta(X)$, $\gamma(\nu^1, \nu^2) \subseteq X$. Otherwise, *some* trajectories which start close to $\Delta(X)$ would evolve in the direction of pure strategies which do not belong to $X$, i.e. away from $\Delta(X)$. When such inclusion applies, the set $X$ is said to be *closed under $\gamma$*. In fact, the next result establishes that such $\gamma$-closedness is not only necessary, but also a sufficient condition, for asymptotic stability.

**Theorem 13 (Ritzberger and Weibull (1995))** *Consider an arbitrary SPES and any $X \subseteq S^1 \times S^2$. The set $\Delta(X)$ is asymptotically stable if, and only if, $X$ is closed under $\gamma$.*

**Proof.** To show sufficiency first, assume $X$ is closed under $\gamma$. Since $\gamma$ is an upper hemi-continuous correspondence (i.e. has a closed graph), there must exist some $\epsilon > 0$ and some associated open neighbourhood of $\Delta(X)$

$$\mathcal{B}(\epsilon) \equiv \{ (\nu^1, \nu^2) \in \Omega : \max [\nu_i^k : k = 1, 2; \ s_i^k \notin X^k] < \epsilon \}$$

such that $\gamma(\mathcal{B}(\epsilon)) \subseteq X$. Otherwise, one could construct a convergent sequence $\{\epsilon_r, \nu_r, y_r\}_{r=1}^\infty$ such that $\nu_r \in \mathcal{B}(\epsilon_r)$, $y_r \in \gamma(\nu_r)$, $y_r \notin X$, for all $r = 1, 2, ...$, with $\epsilon_r \to 0$ and $\nu_r \to \nu^* \in \Delta(X)$, $y_r \to y^*$. Since the set $S \backslash X$ is closed,[6] we should have $y^* \notin X$, which contradicts the upper hemi-continuity of $\gamma$.

Having established the existence of some such neighbourhood $\mathcal{B}(\epsilon)$, suppose that $(\nu^1(t), \nu^2(t)) \in \mathcal{B}(\epsilon)$ at some given $t$. Since the evolutionary system is assumed sign-preserving, it must be that $\dot{\nu}_i^k < 0$ for all $s_i^k \notin X^k$ and each $k = 1, 2$. Thus, the ensuing trajectory never leaves $\mathcal{B}(\epsilon)$, eventually converging to $\Delta(X)$.

---

[6] As is customary, the notation $\backslash$ represents set substraction, i.e. $S \backslash X = \{(s_i^1, s_j^2) \in S : (s_i^1, s_j^2) \notin X\}$.

For the necessity part of the theorem, suppose that $X$ is *not* closed under $\gamma$. That is, there is some $k \in \{1, 2\}$, $s_i^k \notin X^k$, $(\nu^1, \nu^2) \in \Delta(X)$, such that

$$\pi^k(s_i^k, \nu^{k'}) \geq \pi^k(\nu^k, \nu^{k'}).$$

Since $\pi^k(\nu^k, \nu^{k'})$ is just a convex combination of pay-offs $\pi^k(s_j^k, s_r^{k'})$ for pure strategies $s_j^k \in X^k$ and $s_r^{k'} \in X^{k'}$, the previous inequality implies that there must exist some such pure strategies for which

$$\pi^k(s_i^k, s_r^{k'}) \geq \pi^k(s_j^k, s_r^{k'}), \tag{4.8}$$

where, recall, $s_i^k \notin X^k$. Consider now initial conditions $(\nu^1(0), \nu^2(0))$ such that $\nu_i^k(0) = \eta$, $\nu_j^k(0) = 1 - \eta$, and $\nu_r^{k'}(0) = 1$ for any arbitrary $\eta > 0$. By (4.8), since the evolutionary system is sign-preserving and no extinct strategies ever arise along the process, it follows that $\dot{\nu}_i^k(t) \geq 0$ and, therefore, $\nu_i^k(t) \geq \eta$ for all $t$, showing the asymptotic *instability* of $\Delta(X)$. This completes the proof of the theorem. ∎

To illustrate the potential usefulness of the characterization provided by the former result, consider a simple two-stage game proposed by van Damme (1989) to motivate issues of forward induction. (This game is also considered by Ritzberger and Weibull (1995).) Individuals of two populations meet at random to play a co-ordination game with the following pay-off table:

|   | A | B |
|---|---|---|
| A | 3, 1 | 0, 0 |
| B | 0, 0 | 1, 3 |

Table 9

However, before entering this (simultaneous) game, the player of population 1 has the option of choosing an *outside option* $R$ which guarantees him a pay-off of 2, providing the player of population 2 with a pay-off of 5. If we now consider the full two-stage game, its reduced normal form is given by the following pay-off matrix:

| 1 \ 2 | A | B |
|---|---|---|
| R | 2, 5 | 2, 5 |
| A | 3, 1 | 0, 0 |
| B | 0, 0 | 1, 3 |

Table 10

The normal-form game described by the previous pay-off table has two Nash equilibria, both of them subgame-perfect in the underlying extensive form. One of them, $(R, B)$, has player 1 choosing the outside option, under the "threat" that player 2 will aim for the worst equilibrium for player 1 if the co-ordination subgame were played. This equilibrium is subgame-perfect but fails to satisfy a criterion of forward induction. This would require that player 2 interprets a move by player 1 into the co-ordination sub-game as a signal that he intends to play the only ensuing equilibrium strategy which is not dominated by his outside option, i.e. strategy $A$. Consequently, the best response by player 2 to this "signal" should be to play $A$ himself, which only reinforces the reason for player 1 to move into the co-ordination subgame *and* play $A$. Thus, $(A, A)$ is the only equilibrium (also subgame-perfect) which doesn't violate forward-induction rationality.

These considerations are by now very familiar to game theorists. For our purposes, the interesting point arising in this respect is that an identical outcome is obtained if the quite involved chain of signalling outlined above is replaced by mere evolutionary arguments. This idea will be explored in more detail in Sub-section 6.5.2 within a stochastic evolutionary scenario. Here, as subsequently explained, it may be derived from the simple requirement of evolutionary stability in the context of *any* SPES.

By Theorem 13, the asymptotically stable sets induced by any SPES can be associated with the product sets of strategies which are closed under the better-response correspondence. It is easy to see that only two such sets exist in the present example. One of them is the (uninteresting) set consisting of *all* the strategies available to each population, i.e. $\hat{X} = \{R, A, B\} \times \{A, B\}$. In general, of course, the interesting sets are those which are minimal with respect to the contemplated closedness requirement. This is indeed the case for the second set which satisfies it, $\tilde{X} = \{A\} \times \{A\}$. As advanced, this singleton coincides with the outcome derived from the usual forward-induction arguments.

### 4.4.3    Long-run regularities

Even if the evolutionary process is not convergent, it may still display significant long-run regularities, e.g. the constancy of some appropriate magnitude over time. As suggested by the proof of Theorem 12, "volume", for an appropriate transformation of the RD, is always one of them. This, however, is not very informative *per se* about any interesting feature of the process. Much more useful would be to establish the constancy of some real function (preferably smooth) of the state space along any dynamic path. Such a function, usually called a *constant of motion*, guarantees that every trajectory of the system remains within some given submanifold of $\Omega$,[7] thus providing some extent of

---

[7] Heuristically, a submanifold of $\Omega$ is just a lower-dimensional surface included in it.

"order" for its long-run behaviour.

But even this, in general, will not be easy to obtain. However, if the system is especially simple, i.e. bidimensional (with two strategies per population), a general and clear-cut result can be obtained for the RD if the underlying game has an interior Nash equilibrium. This conclusion is established by the following result, Section 4.7 providing some illustrative examples.

**Theorem 14 (Hofbauer and Sigmund (1988))** *Let $A, B \in \Re^{2 \times 2}$ and assume the corresponding bimatrix game has an interior Nash equilibrium $\hat{\sigma} \in \Sigma$. (That is, $\hat{\sigma}_i^k > 0$, $\forall k = 1, 2$, $\forall i = 1, 2$.) Then, generically, there is some $c \in \Re$ such that $\zeta(\nu) = c \sum_{i=1,2} \hat{\sigma}_i^1 \log \nu_i^1 - \sum_{i=1,2} \hat{\sigma}_i^2 \log \nu_i^2$ is a constant of motion for the RD.*

**Proof.** The first point to note is that, with only two strategies for each population, it is generically possible to find real values $\alpha_1, \alpha_2, \beta_1, \beta_2$, and $c$ such that

$$\tilde{a}_{ij} = a_{ij} + \alpha_j, \quad \tilde{b}_{ij} = b_{ij} + \beta_i, \quad i, j = 1, 2;$$

$$c\,\tilde{A} = \tilde{B}, \tag{4.9}$$

where $\tilde{A} = (\tilde{a}_{ij})_{i,j=1,2}$; $\tilde{B} = \left(\tilde{b}_{ij}\right)_{i,j=1,2}$. (Here, the only point to check is that the corresponding linear system of equations has a solution unless the original pay-off matrices exhibit some precise "ties" among their entries.)

Given the function $\zeta(\cdot)$ defined in the statement of the Theorem, it follows that:

$$\begin{aligned} \dot{\zeta}(\nu) &= c \sum_{i=1,2} \hat{\sigma}_i^1 \frac{\dot{\nu}_i^1}{\nu_i^1} - \sum_{i=1,2} \hat{\sigma}_i^2 \frac{\dot{\nu}_i^2}{\nu_i^2} \\ &= c\left(\hat{\sigma}^1 - \nu^1\right) \cdot \tilde{A}\nu^2 - \nu^1 \cdot \tilde{B}\left(\hat{\sigma}^2 - \nu^2\right), \end{aligned} \tag{4.10}$$

where one relies on the additive invariance of the RD (cf. Subsection 3.2.3). From the fact that, at an interior Nash equilibrium, all pure strategies must earn the same pay-off, it follows that:

$$\left(\hat{\sigma}^1 - \nu^1\right) \cdot \tilde{A}\hat{\sigma}^2 = \hat{\sigma}^1 \cdot \tilde{B}\left(\hat{\sigma}^2 - \nu^2\right) = 0,$$

which allows (4.10) to be rewritten as follows:

$$\dot{\zeta}(\nu) = c\left(\hat{\sigma}^1 - \nu^1\right) \cdot \tilde{A}\left(\nu^2 - \hat{\sigma}^2\right) - \left(\nu^1 - \hat{\sigma}^1\right) \cdot \tilde{B}\left(\hat{\sigma}^2 - \nu^2\right),$$

which, by virtue of (4.9), is identically equal to zero. This completes the proof of the Theorem. ∎

## 4.5     Evolution and Rationality

Even if, in general, one may not rely on the stability of equilibria in order to obtain some clear-cut dynamic prediction from an evolutionary model, some basic questions about its long-run implications may still be asked. For example, there is the important issue of whether evolution will be able to narrow down the span of admissible strategies in the long run, discarding all suboptimal behaviour. In fact, it has been precisely a positive, but merely heuristic, answer to this question which has underlain much of the *"as-if"* motivations of (unbounded) rationality often found in traditional economic theory.

To approach this issue in a rigorous fashion, the first question to settle is what criterion for suboptimal (or "irrational") behaviour one should consider. In this respect, the elimination of dominated strategies seems to represent a minimal demand of rationality. However, as readers familiar with Game Theory will be well aware, such domination can be formulated in a variety of different ways:

1. Strict domination in terms of pure strategies alone,
2. Strict domination in terms of arbitrary, possibly mixed, strategies,
3. Weak domination, again in pure or mixed strategies.

As we shall see, our conclusions will significantly differ depending on the particular concept under consideration.

Once this issue is settled, a natural second step to take involves exploring whether evolutionary dynamics might be responsive to a *repeated* application of any of the above domination criteria. This, for our two-population scenario and criterion 2 above, leads to the familiar notion of *rationalizability* (Bernhein (1984) and Pearce (1984)). This concept, it is often argued, embodies the essential idea of rationality that (minimally or maximally, depending on one's particular emphasis) game-theoretic analysis should insist upon.

### 4.5.1     Evolution and pay-off dominance

First, the three different criteria of domination to be considered here are formally defined.

**Definition 12** *For each population $k = 1, 2$, strategy $s_i^k \in S^k$ is said to be dominated in pure strategies if $\exists s_j^k \in S^k$ such that:*

$$\forall s_r^{k'} \in S^{k'}, \quad \pi^k\left(s_j^k, s_r^{k'}\right) > \pi^k\left(s_i^k, s_r^{k'}\right).$$

**Definition 13** *For each population $k = 1, 2$, strategy $s_i^k \in S^k$ is said to be dominated (in mixed strategies) if $\exists \sigma^k \in \Sigma^k$ such that:*

$$\forall s_r^{k'} \in S^{k'}, \quad \pi^k\left(\sigma^k, s_r^{k'}\right) > \pi^k\left(s_i^k, s_r^{k'}\right).$$

**Definition 14** *For each population* $k = 1, 2$, *strategy* $s_i^k \in S^k$ *is said to be weakly dominated if* $\exists \sigma^k \in \Sigma^k$ *such that:*

$$\forall s_r^{k'} \in S^{k'}, \quad \pi^k \left( \sigma^k, s_r^{k'} \right) \geq \pi^k \left( s_i^k, s_r^{k'} \right);$$

$$\exists s_{r_o}^{k'} \in S^{k'}, \quad \pi^k \left( \sigma^k, s_{r_o}^{k'} \right) > \pi^k \left( s_i^k, s_{r_o}^{k'} \right).$$

Denote by $D^k$, $\hat{D}^k$, and $\tilde{D}^k$ the strategies of population $k$ which are dominated in pure strategies, in mixed strategies, and weakly dominated respectively. Obviously, $D^k \subseteq \hat{D}^k \subseteq \tilde{D}^k$.

The first (and weakest) conclusion concerns the strongest domination criterion, i.e. that involving only pure strategies.

**Theorem 15 (Samuelson and Zhang (1992))** *Let* $s_i^k \in D^k$ *for some* $k = 1, 2$. *Then, every trajectory* $\nu(\cdot)$ *of a GM evolutionary system with* $\nu(0) \in int(\Omega)$ *satisfies* $\lim_{t \to \infty} \nu_i^k(t) = 0$.

**Proof.** Let $s_i^k \in D^k$ for some $k = 1, 2$, and let $\varepsilon > 0$ be chosen such that for some $s_j^k \in S^k$, [8]

$$\forall \nu^{k'} \in \Delta^{m_{k'}-1}, \quad \pi^k \left( s_j^k, \nu^{k'} \right) - \pi^k \left( s_i^k, \nu^{k'} \right) \geq \varepsilon.$$

Such an $\varepsilon$ always exists due to the continuity of each $\pi^k(\cdot)$ and the compactness of $\Delta^{m_{k'}-1}$. Thus, from the assumed growth-monotonicity (together with continuity) of the evolutionary system, there is some $\delta > 0$ such that:

$$\frac{\dot{\nu}_i^k(t)}{\nu_i^k(t)} - \frac{\dot{\nu}_j^k(t)}{\nu_j^k(t)} \leq -\delta$$

for all $t$, which implies that

$$\frac{\nu_i^k(t)}{\nu_j^k(t)} \leq \frac{\nu_i^k(0)}{\nu_j^k(0)} e^{-\delta t},$$

thus proving the desired conclusion. ∎

Suppose now that, in line with the domination criterion presented in Definition 13, the issue becomes whether all pure strategies that are (strictly) dominated by mixed ones are certain to be played in vanishing long-run frequencies. One may construct examples (see Bjönerstedt *et al.* (1993)) which indicate that the monotonicity requirements proposed in Subsection 4.3.2 need to be substantially strengthened. For example, if one is willing to restrict to the Replicator Dynamics, the next result guarantees the desired conclusion.

---

[8] Note, of course, that if $s_j^k$ dominates $s_i^k$ when the opponent plays *any* given pure strategy (cf. Definition 12), this also happens when the particular strategy to be faced is determined by any given probability vector $\nu^{k'}$.

**Theorem 16 (Samuelson and Zhang (1992))** *Let $s_i^k \in \hat{D}^k$ for some $k = 1, 2$. Then, every trajectory $\nu(\cdot)$ of the Replicator Dynamics with $\nu(0) \in int(\Omega)$ satisfies $\lim_{t \to \infty} \nu_i^k(t) = 0$.*

**Proof.** Let $s_i^k \in \hat{D}^k$ for some $k = 1, 2$, and let $\varepsilon > 0$ be chosen such that, for some $\sigma^k \in \Sigma^k$, one has:

$$\forall \nu^{k'} \in \Delta^{m_{k'}-1}, \quad \pi^k\left(\sigma^k, \nu^{k'}\right) - \pi^k\left(s_i^k, \nu^{k'}\right) \geq \varepsilon. \tag{4.11}$$

Again, such an $\varepsilon$ always exists due to the continuity of each $\pi^k(\cdot)$ and the compactness of $\Delta^{m_{k'}-1}$. Define the function $W : \Delta^{m_k-1} \to \Re$ as follows:

$$W\left(\nu^k\right) = \sum_{j=1}^{m_k} \sigma_j^k \cdot \log \nu_j^k - \log \nu_i^k.$$

Along any trajectory of the RD which starts in the interior of $\Omega$, its time derivative may be computed as follows:

$$
\begin{aligned}
\dot{W}\left(\nu^1(t), \nu^2(t)\right) &= \sum_{j=1}^{m_k} \sigma_j^k \frac{\dot{\nu}_j^k(t)}{\nu_j^k(t)} - \frac{\dot{\nu}_i^k(t)}{\nu_i^k(t)} \\
&= \sum_{j=1}^{m_k} \sigma_j^k \left[ \begin{array}{c} \pi^k\left(s_j^k, \nu^{k'}(t)\right) - \\ \sum_{r=1}^{m_k} \nu_r^k(t) \pi^k\left(s_r^k, \nu^{k'}(t)\right) \end{array} \right] \\
&\quad - \left[ \pi^k\left(s_i^k, \nu^{k'}(t)\right) - \sum_{j=1}^{m_k} \nu_j^k(t) \pi^k\left(s_j^k, \nu^{k'}(t)\right) \right] \\
&= \sum_{j=1}^{m_k} \sigma_j^k \pi^k\left(s_j^k, \nu^{k'}(t)\right) - \pi^k\left(s_i^k, \nu^{k'}(t)\right)
\end{aligned}
$$

which, by (4.11) is bounded below by $\varepsilon$. Therefore,

$$\lim_{t \to \infty} W\left(\nu^1(t), \nu^2(t)\right) = \infty$$

which implies, as desired, that $\lim_{t \to \infty} \nu_i^k = 0$. ∎

There are two significant elaborations on the previous result which are worth summarizing.

(i)   Samuelson and Zhang (1992) show that, in fact, the conclusion of Theorem 16 follows from a weaker hypothesis on the evolutionary system. Specifically, the system simply needs to be *aggregate monotonic*, as described by the following definition.

**Definition 15** *An evolutionary system (4.1) is said to be aggregate-monotonic if $\forall k = 1, 2, \ \forall \sigma^k, \hat{\sigma}^k \in \Sigma^k, \forall \nu \in \Omega$,*

$$\pi^k\left(\sigma^k, \nu^{k'}\right) \geq \pi^k\left(\hat{\sigma}^k, \nu^{k'}\right) \Leftrightarrow \sigma^k \cdot F^k(\nu) \geq \hat{\sigma}^k \cdot F^k(\nu).$$

The concept of aggregate-monotonicity demands for mixed strategies the counterpart of the monotonicity condition contemplated by Definition 10 for pure strategies. Obviously, since a pure strategy is just a degenerate mixed strategy, growth monotonicity is weaker than aggregate monotonicity. Whereas the former allows the frequencies of pure strategies to change in any arbitrary manner as long as they respect the (ordinal) ranking contemplated by Definition 10, the latter introduces a cardinal requirement on their respective rates of change. It requires that, even though the system is defined on pure-strategy frequencies, it should always evolve in the direction of higher-pay-off mixed strategies.

Samuelson and Zhang (1992, Theorem 3) show that the aggregate-monotonicity condition is closely related to the Replicator Dynamics. Specifically, every aggregate monotonic system represents a "variable scaling" of the RD which may be written as follows:

$$\dot{\nu}_i^k(t) = \lambda^k(\nu(t)) \, \nu_i^k(t) \left[ \pi^k\left(s_i^k, \nu^{k'}(t)\right) - \sum_{j=1}^{m_k} \nu_j^k(t) \, \pi^k\left(s_j^k, \nu^{k'}(t)\right) \right],$$
(4.12)

for each $i = 1, 2, ..., m_k$; $k, k' = 1, 2$ $(k \neq k')$, where the factor $\lambda^k(\cdot) > 0$ in the above expression *only* depends on the population $k = 1, 2$ under consideration, *not* on the particular strategy. If $\lambda^1(\cdot) \equiv \lambda^2(\cdot) \equiv 1$, we obtain the RD (recall Subsection 4.3.3). Slightly more generally, if $\lambda^1(\cdot) \equiv \lambda^2(\cdot)$, the system is essentially equivalent to the RD since, as explained above, it only differs from it in that a variable rate of adjustment is allowed.

(ii) Theorem 16 crucially depends on the continuous-time formalization of the RD. Dekel and Scotchmer (1992) have shown that, in a discrete-time version of the RD, a dominated strategy (in the sense of Definition 13) may survive in the long run. However, Cabrales and Sobel (1992) make it clear that this depends on the adjustment framework *not* being sufficiently gradual. If it is gradual enough, only non-dominated strategies may persist in the long run.

Let us now turn to the third and weakest criterion of domination presented in Definition 14. It is not difficult to find particular scenarios where evolutionary processes do *not* weed out weakly dominated strategies for *any* (interior) initial conditions. Since this claim is of a negative nature, it becomes stronger the more demanding is the evolutionary system under consideration. For example, Samuelson and Zhang (1992) consider the Replicator Dynamics[9] applied to a bilateral game with pay-off matrices

$$A = \begin{bmatrix} 1 & 1 \\ 1 & 0 \end{bmatrix}; \quad B = \begin{bmatrix} 1 & 0 \\ 1 & 0 \end{bmatrix}.$$

---

[9] In this respect, recall that the RD not only satisfies both of the monotonicity criteria proposed in Subsection 4.3.2 but, in view of Theorem 16 and point (i) above, may be seen as a strong selection mechanism.

If one denotes $x(t) \equiv \nu_1^1(t),$ and $y(t) \equiv \nu_1^2(t)$, the RD applied to the example induces the following system of differential equations:

$$\begin{aligned} \dot{x}(t) &= x(t)\ (1 - x(t))(1 - y(t)) \\ \dot{y}(t) &= y(t)\ (1 - y(t)). \end{aligned} \qquad (4.13)$$

Assume interior initial conditions $x(0), y(0) \in (0,1)$ and let $z(t) \equiv \frac{1-x(t)}{y(t)}$. Then,

$$\frac{\dot{z}(t)}{z(t)} = -(1 + x(t))\ (1 - y(t))$$

and, therefore,

$$z(t) = z(0)\ e^{-\int_{\tau=0}^{t}(1+x(\tau))(1-y(\tau))d\tau}.$$

Since, from (4.13),

$$\int_{\tau=0}^{t} (1 + x(\tau))\ (1 - y(\tau))\ d\tau \leq 2(1 - y(0)) \int_{\tau=0}^{t} e^{-y(0)\cdot\tau}d\tau < +\infty$$

we have

$$\lim_{t\to\infty} z(t) > 0,$$

which indicates that the frequency of individuals of population 1 adopting the weakly dominated strategy must remain positive in the long run for all interior initial conditions.

To summarize what has been presented so far, evolutionary forces have been shown to be responsive to considerations of (strict) domination. Whereas those based only on pure strategies are reflected by any evolutionary system displaying growth-monotonicity, the weaker notion based on mixed strategies requires an evolutionary system which is a close "relative" of the RD. With respect to weak (i.e. non-strict) concepts of domination, evolutionary systems, even demanding ones, fare much worse. In general, not even the RD guarantees that a weakly dominated strategy is not played in the long run in positive frequency. Thus, in particular, as in the previous example, it may induce a limit state which corresponds to a Nash equilibrium which is *not* perfect.[10]

### 4.5.2    *Evolution, iterative dominance, and rationalizability*

As straightforward corollaries of Theorems 15 and 16, it is now verified that their conclusions apply not just to a strategy which is dominated when *all* possible strategies are available, but also to any strategy which becomes dominated only

---

[10] Recall that, as explained in the proof of Proposition 2, if an equilibrium is perfect, it *cannot* involve playing weakly dominated strategies with positive probability.

after repeated application of the corresponding dominance criterion. This leads us to two alternative notions of iterative dominance, which are formalized as follows.

**Definition 16** *For each population* $k = 1, 2$, *define iteratively the following sets:*

$$S^{k0} = S, \quad D^{k0} = D^k,$$

*and, for* $q = 1, 2, ...,$

$$S^{kq} = S^{k(q-1)} \backslash D^{k(q-1)},$$
$$D^{kq} = \{s_i^k \in S^{kq} : \exists s_j^k \in S^{kq} \text{ s. t. } \forall s_r^{k'} \in S^{k'q}, \pi^k \left( s_j^k, s_r^{k'} \right) > \pi^k \left( s_i^k, s_r^{k'} \right) \}.$$

*The set* $S^{k*} \equiv \bigcap_{q=0}^{\infty} S^{kq}$ *is called the set of strongly[11] iteratively undominated strategies of population* $k$.

**Definition 17** *For each population* $k = 1, 2$, *define iteratively the following sets:*

$$\hat{S}^{k0} = S^k; \hat{D}^{k0} = \hat{D}^k;$$

*and for* $q = 1, 2, ...,$

$$\hat{S}^{kq} = \hat{S}^{k(q-1)} \backslash \hat{D}^{k(q-1)},$$
$$\hat{\Sigma}^{qk} = \{\sigma^k \in \Sigma^k : \sigma_i^k = 0 \text{ for } i \notin \hat{S}^{kq}\},$$
$$\hat{D}^{kq} = \left\{ \begin{array}{c} s_i^k \in \hat{S}^{kq} : \exists \sigma^k \in \hat{\Sigma}^{qk} \text{ s. t.} \\ \forall s_r^{k'} \in \hat{S}^{k'q}, \pi^k \left( \sigma^k, s_r^{k'} \right) > \pi^k \left( s_i^k, s_r^{k'} \right) \end{array} \right\}.$$

*The set* $\hat{S}^{k*} \equiv \bigcap_{q=0}^{\infty} \hat{S}^{kq}$ *is called the set of iteratively undominated (or rationalizable)[12] strategies of population* $k$.

The following two corollaries of Theorems 15 and 16 formally state the advanced conclusions.

**Corollary 3** *Let* $s_i^k \notin S^{k*}$, $k = 1, 2$. *Then, every trajectory* $\nu(\cdot)$ *of a GM evolutionary system with* $\nu(0) \in int(\Omega)$ *satisfies* $\lim_{t \to \infty} \nu_i^k(t) = 0$.

---

[11] Here, the term "strongly" refers to the fact that the considered concept of dominance in pure strategies is the strongest one proposed; reciprocally, of course, the corresponding concept of undomination is the weakest. Incidentally, note that, by construction, the sets $S^{k*}$ (as well as the sets $\hat{S}^{k*}$ in Definition 17 below) are always non-empty.

[12] It is well known that, for *bilateral* games, the set of iteratively undominated strategies coincides with the set of rationalizable strategies (see Bernheim (1984), Pearce (1984)). Being standard, an independent general definition for this concept is not provided here. As shown by Brandemburger and Dekel (1987), one can associate rationalizability with underlying common knowledge of rationality. It is in this sense that, as mentioned, one may informally think of rationalizability as representing the core implication of rationality in the analysis of games.

**Corollary 4** *Let $s_i^k \notin \hat{S}^{k*}$, $k = 1, 2$. Then, every trajectory $\nu(\cdot)$ of the Replicator Dynamics with $\nu(0) \in int(\Omega)$ satisfies $\lim_{t \to \infty} \nu_i^k(t) = 0$.*

The proof of the above corollaries is an immediate consequence of the line of argument used to prove the theorems above. Simply note that any strategies which would become dominated if some other strategies were discarded, will also appear dominated if these strategies are present in a sufficiently small frequency. (Here, of course, one must rely on the continuity of the pay-off functions.) But then, if these other strategies are dominated in the first place, their frequency will become arbitrarily small once the evolutionary system has proceeded for long enough, by virtue of Theorems 15 or 16 (whichever applies for the domination criterion in question). The inductive reasoning underlying the previous corollaries then becomes apparent. Applied to any arbitrary level, it implies that the consequences (as well as the caveats) which evolution was seen to have on "single-level rationality" (recall the preceding subsection) extend to the multi-level, hierarchic considerations arising here.

## 4.6    General Evolutionary Processes

The family of evolutionary processes defined in (4.1) is quite restrictive in a number of respects. For example, every extinct strategy is forced to remain so forever since growth *rates* are given by functions $F_i^k(\cdot)$ which are assumed Lipschitz-continuous. A more flexible formulation can bypass this limitation. In general, one may simply postulate an evolutionary system of the form

$$\dot{\nu}_i^k(t) = G_i^k\left(\nu^1(t), \nu^2(t)\right), \quad i = 1, 2, ..., m_k, \ k = 1, 2, \qquad (4.14)$$

for some set of functions $G_i^k : \Omega \to R$. We shall not insist that the functions $G_i^k$ be Lipschitz, or even continuous, on the whole of $\Omega$, now interpreting the dot in (4.14) as right-derivatives when necessary. Of course, it will be assumed that a unique solution to the dynamical system exists. This requires, in particular, that

$$\sum_{i=1}^{m_k} G_i^k(\nu) = 0$$
$$\nu_i^k = 0 \Rightarrow G_i^k(\nu) \geq 0 \qquad (4.15)$$

for each $k = 1, 2$, $i = 1, 2, ..., m_k$.

The above general formulation permits the consideration of a wide range of monotonicity concepts, not necessarily focused on growth rates as in Definitions 10 and 11. Depending on the particular scenario under consideration (see some illustrative examples below), these alternative concepts might be better suited to describe the evolutionary forces at work.

### 4.6.1 *Gradient monotonicity*

As an interesting variation on the general idea of evolutionary monotonicity, consider the following alternative to Definition 10.[13]

**Definition 18** *An evolutionary system (4.14) is said to be gradient-monotonic if* $\forall k = 1, 2, \forall \nu = (\nu^1, \nu^2) \in \Omega, \forall i, j = 1, 2, ..., m_k,$

$$\left[ \pi^k \left( s_i^k, \nu^{k'} \right) > \pi^k \left( s_j^k, \nu^{k'} \right), \nu_j^k > 0 \right] \Rightarrow G_i^k (\nu) > G_j^k (\nu).$$

To understand the previous concept, consider any two strategies $s_i^k$ and $s_j^k$ such that the pay-off for the first is higher than for the second. *Gradient monotonicity* simply requires that the *only* case in which the (absolute) magnitudes of change in their respective frequencies may *not* reflect this difference in pay-offs is when $\nu_j^k = 0$. In these circumstances, one must have $G_j^k (\nu) \geq 0$ due to the boundary constraints which ensure that the system is well defined. However, if $\nu_i^k > 0$, a gradient-monotonic evolutionary system would still find it admissible that, in this case, $G_i^k (\nu) < 0$.

### 4.6.2 *Dynamic stability and rationality*

In Subsection 4.4.1, it was established that every asymptotically stable state of either a growth-monotonic or a sign-preserving evolutionary system is associated with a Nash equilibrium of the underlying game. Now, the result is stronger in that the same conclusion applies to *all* rest points of the dynamics, be they asymptotically stable or not.

**Theorem 17** *Let $\hat{\nu} \equiv (\hat{\nu}^1, \hat{\nu}^2)$ be an equilibrium (or rest point) of a gradient-monotonic evolutionary system (4.14).Then, $(\hat{\nu}^1, \hat{\nu}^2)$, as an element of $\Sigma \equiv \Sigma^1 \times \Sigma^2$, is a Nash equilibrium of the underlying bimatrix game.*

**Proof.** Every rest point $\hat{\nu}$ of the evolutionary system must satisfy:

$$\forall k = 1, 2, \ \forall i = 1, 2, ..., m_k, \ \ G_i^k (\nu) = 0. \tag{4.16}$$

Consider, for each $k = 1, 2$, any strategy $s_r^k \in S^k$ such that $\hat{\nu}_r^k > 0$. By (4.16) and gradient monotonicity, it follows that:

$$\forall s_i^k \in S^k, \ \ \pi^k \left( s_i^k, \hat{\nu}^{k'} \right) \leq \pi^k \left( s_r^k, \hat{\nu}^{k'} \right),$$

which proves the desired conclusion. ∎

---

[13] Although reminiscent of absolute monotonicity in Nachbar (1990), and order compatibility in Friedman (1991), the concept proposed here is quite different. As illustrated below, its implications also differ.

Finally, as a further illustration of the implications of gradient-monotonic systems, we present a counterpart of Theorem 15 above. It requires two further qualifications on the evolutionary system which are contained in the following assumptions:

*Assumption PR* (Pay-off Responsiveness):   There exists some strictly increasing continuous function $\varphi : \Re_+ \to \Re_+$ such that $\forall k = 1, 2$, $\forall \nu \in \Omega, \forall i, j = 1, 2, ..., m_k,$

$$\left[ \pi^k \left( s_i^k, \nu^{k'} \right) > \pi^k \left( s_j^k, \nu^{k'} \right), \; \nu_j^k > 0 \right] \Rightarrow$$
$$G_i^k (\nu) - G_j^k (\nu) \; \geq \; \varphi \left( \pi^k \left( s_i^k, \nu^{k'} \right) > \pi^k \left( s_j^k, \nu^{k'} \right) \right).$$

*Assumption SMB* (Strong Monotonicity at the Boundary):   $\forall k = 1, 2$, $\forall \nu \in \Omega, \; \forall i = 1, 2, ..., m_k,$

$$\left[ \nu_i^k = 0, \; G_i^k (\nu) > 0 \right] \Rightarrow \forall j \neq i, \; \pi^k \left( s_i^k, \nu^{k'} \right) \geq \pi^k \left( s_j^k, \nu^{k'} \right).$$

Assumption PR requires that the system's sensitivity increases continuously in pay-off differences. It prevents, for example, the differential rate of change between two strategies becoming arbitrarily small even though their pay-off difference might remain bounded above zero. Assumption SMB, on the other hand, is a relaxation of the concept of strong monotonicity contemplated in Samuelson (1988) in that it only applies at points on the boundary of $\Omega$. It requires that if a previously extinct strategy is to enter the population, it must not be dominated by any other. Or, in other words, it reflects the idea that no "mutation" will be able to gain a significant foothold in the population if some other strategy is faring better.

The following result establishes that gradient-monotonic systems ensure a degree of long-run rationality comparable to that of GM evolutionary systems:

**Theorem 18** *Let* $s_i^k \in D^k$ *for some* $k = 1, 2$ *(cf. Subsection 4.5.1). Then, every trajectory* $\nu (\cdot)$ *of a gradient-monotonic evolutionary system (4.14) which satisfies Assumptions PR and SMB induces* $\lim_{t \to \infty} \nu_i^k (t) = 0$.

**Proof.** Let $s_i^k \in D^k$ for some $k = 1, 2$. Of course, if $\nu_i^k (0) = 0$, then $\nu_i^k (t) = 0$ for all $t \geq 0$ due to Assumption SMB. Thus, assume $\nu_i^k (0) > 0$.

The continuity of the pay-off functions and the compactness of $\Omega$ imply that $\exists s_j^k \in S^k$ such that

$$\pi^k \left( s_j^k, \nu^{k'} \right) - \pi^k \left( s_i^k, \nu^{k'} \right) \geq \xi$$

for some $\xi > 0$. Therefore, by gradient monotonicity of the evolutionary system and Assumption PR, it follows that, for any $t$,

$$\nu_i^k(t) > 0 \Rightarrow \dot{\nu}_i^k(t) - \dot{\nu}_j^k(t) \leq -\varphi(\xi).$$

Hence, for every $\tau \geq 0$,

$$[\nu_i(t) > 0, \; \forall t \leq \tau] \Rightarrow \int_{t=0}^{\tau} \dot{\nu}_i^k(t) \; dt \leq -\varphi(\xi)\tau + \int_{t=0}^{\tau} \dot{\nu}_j^k(t) \; dt.$$

Therefore,

$$
\begin{aligned}
\nu_i^k(\tau) &= \nu_i^k(0) + \int_{t=0}^{\tau} \dot{\nu}_i^k(t) \; dt \\
&\leq \nu_i^k(0) - \varphi(\xi)\tau + \int_{t=0}^{\tau} \dot{\nu}_j^k(t) \; dt \\
&= \nu_i^k(0) - \varphi(\xi)\tau + \nu_j^k(\tau) - \nu_j^k(0) \\
&\leq 2 - \varphi(\xi)\tau.
\end{aligned}
$$

Thus, for some $\tau' \in [0, 2/\varphi(\xi)]$, $\nu_i^k(\tau') = 0$. Beyond this point, i.e. for all $\tau'' \geq \tau'$, $\nu_i^k(\tau'') = 0$, again by Assumption SMB. ∎

## 4.7 Examples

Two different types of scenarios will be considered here, which are simple il-lustrations of the general random-matching framework proposed in Section 4.2. In the first of them, interaction between populations exhibits strategic comple-mentarities. In contrast, the second scenario displays a highly competitive (in fact, zero-sum) strategic interaction.

### 4.7.1 Trading complementarities

Consider the following context, reminiscent of the well-known work of Diamond (1982). There are two islands, 1 and 2. At every point in time $t$, each agent of island $k = 1, 2$ is in one of two states: employed or unemployed. In the former case, he produces $y > 0$ units of consumption good and incurs a utility cost of $c > 0$ $(c < y)$; in the latter case, he simply produces no output and incurs no cost.

Assume that the consumption good produced in one island cannot be prof-itably consumed by agents of the same island. In order to derive utility from it, the good needs to be exchanged for that produced in the other island. More specifically, suppose that at every $t \geq 0$, agents of both islands are randomly matched in pairs. If two *employed* individuals meet, they exchange their goods and they both enjoy a utility of $y - c$. Otherwise, no trade takes place and the pay-off is zero if the worker is unemployed and $-c$ if he is employed (but cannot exchange his produced good).

If we identify $s_1^k$ with being unemployed and $s_2^k$ with being employed in island $k$, their corresponding expected pay-offs for a population state $\nu = ((\nu_1^1, \nu_2^1), (\nu_1^2, \nu_2^2))$ are as follows:

$$\begin{aligned} \pi^k(s_1^k, \nu^{k'}) &= 0 \\ \pi^k(s_2^k, \nu^{k'}) &= (-c) \cdot \nu_1^{k'} + (y - c) \cdot \nu_2^{k'}. \end{aligned}$$

Consider now *any* evolutionary system (either of the form (4.1) or (4.14)) satisfying one of the monotonicity concepts proposed. It can be seen at once that the population states

$$\begin{aligned} \hat{\nu} &= ((1,0), (1,0)) \\ \tilde{\nu} &= ((0,1), (0,1)), \end{aligned}$$

where either nobody is employed in the two islands (state $\hat{\nu}$) or everyone is in both of them (state $\tilde{\nu}$) are asymptotically stable. Moreover, there is a population state $\nu^*$ defined by:

$$\nu^{k*} = \left(1 - \frac{c}{y}, \frac{c}{y}\right); \quad k = 1, 2,$$

such that

$$\left[\nu_1^k(0) \geq 1 - \frac{c}{y}, \ k = 1, 2, \ \nu(0) \neq \nu^*\right] \Rightarrow \lim_{t \to \infty} \nu(t) = \hat{\nu}, \qquad (4.17)$$

$$\left[\nu_1^k(0) \leq 1 - \frac{c}{y}, \ k = 1, 2, \ \nu(0) \neq \nu^*\right] \Rightarrow \lim_{t \to \infty} \nu(t) = \tilde{\nu}. \qquad (4.18)$$

All other initial conditions for which neither (4.17) nor (4.18) apply belong to the basin of attraction of one of the states $\hat{\nu}$ or $\tilde{\nu}$, except for a one-dimensional manifold (of measure zero) which goes through $\nu^*$ and separates both of them, its particular form depending on the details of the evolutionary system considered. Figure 5 illustrates the situation.

### 4.7.2    Risky trading

Consider again a trading context between two populations, 1 and 2, whose members are randomly matched in pairs (one from each population) every period. All individuals produce *two* units of a certain *population-specific* commodity. However, to be enjoyed, this commodity has to be consumed together with that produced by the other population.

The two populations are *not* symmetric. Population 1 is potentially aggressive. A certain fraction $\nu_1^1$ of them, when matched with an individual of population 2, will attempt to forcefully deprive the latter of his two commodity

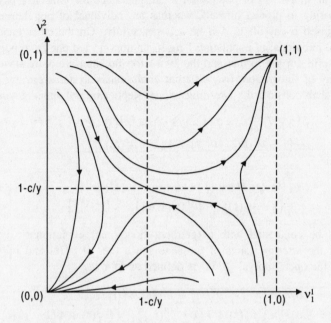

Figure 5: Trading complementarities, a monotonic evolutionary system

units (strategy $s_1^1$). On the other hand, the complementary fraction $\nu_2^1 = 1 - \nu_1^1$ are peaceful individuals (they adopt strategy $s_2^1$) and exchange one unit of their produced commodity for a corresponding unit of the commodity produced by the other population.

Faced with this state of affairs, individuals of population 2 can adopt one of two possible strategies. They can either obtain protection at some cost (strategy $s_1^2$), or instead risk unprotected trading (strategy $s_2^2$). The fractions of the population which decide on each of these options are denoted by $\nu_1^2$ and $\nu_2^2$, respectively.

Motivated by the previous story, the following pay-off table is postulated:

| 2<br>1 | $s_1^2$ | $s_2^2$ |
|---|---|---|
| $s_1^1$ | 1, 1 | 7, −2 |
| $s_2^1$ | 4, 1 | 4, 4 |

Table 11

Thus, an individual of population 2 can guarantee for himself a pay-off of 1 by choosing to protect himself, whereas an individual of population 1 can ensure himself a pay-off of 4 if he acts peacefully. On the other hand, if an aggressive individual of population 1 meets an unprotected one from population 2, the benefits for the former and the losses for the latter are both substantial.

By way of illustration, two different evolutionary systems are considered. First, we shall consider the two-dimensional Replicator Dynamics given by:[14]

$$
\dot{\nu}_1^1(t) = \nu_1^1(t) \left\{ \nu_1^2(t) + 7 \left(1 - \nu_1^2(t)\right) - \left[ \nu_1^1(t) \left( \nu_1^2(t) + 7 \left(1 - \nu_1^2(t)\right) \right) \right. \right.
$$
$$
\left. \left. + \left(1 - \nu_1^1(t)\right) \left( 4\nu_1^2(t) + 4 \left(1 - \nu_1^2(t)\right) \right) \right] \right\}
\tag{4.19}
$$

$$
\dot{\nu}_1^2(t) = \nu_1^2(t) \left\{ \nu_1^1(t) + \left(1 - \nu_1^1(t)\right) - \left[ \nu_1^2(t) \left( \nu_1^1(t) + \left(1 - \nu_1^1(t)\right) \right) \right. \right.
$$
$$
\left. \left. + \left(1 - \nu_1^2(t)\right) \left( -2\nu_1^1(t) + 4 \left(1 - \nu_1^1(t)\right) \right) \right] \right\}.
\tag{4.20}
$$

It will be contrasted with a (gradient-monotonic) evolutionary dynamics which for the *interior* points of the state space (i.e. if $\nu_1^1(t)$ and $\nu_1^2(t)$ both belong to the open interval $(0, 1)$) is defined as follows:

$$
\dot{\nu}_1^1(t) = \alpha \, sgn \left[ \pi^1 \left( s_1^1, \nu^2(t) \right) - \pi^1 \left( s_2^1, \nu^2(t) \right) \right]
\tag{4.21}
$$
$$
= \alpha \, sgn \left[ \left( \nu_1^2(t) + 7 \left(1 - \nu_1^2(t)\right) \right) - \left( 4\nu_1^2(t) + 4 \left(1 - \nu_1^2(t)\right) \right) \right].
$$

$$
\dot{\nu}_1^2(t) = \beta \, sgn \left[ \left( \pi^2 \left( s_1^2, \nu^1(t) \right) - \pi^2 \left( s_2^2, \nu^1(t) \right) \right) \right]
\tag{4.22}
$$
$$
= \beta \, sgn \left[ \left( \nu_1^1(t) + \left(1 - \nu_1^1(t)\right) \right) - \left( -2\nu_1^1(t) + 4 \left(1 - \nu_1^1(t)\right) \right) \right],
$$

where $\alpha, \beta > 0$ are interpreted as the rate at which population 1 or 2 respectively adjust in a pay-off-monotone direction. The system is then extended to the boundary of the state space as follows. Let $\varphi^1(\nu)$ and $\varphi^2(\nu)$ represent the functions which define the RHS of expressions (4.21) and (4.22). When the current state is on the boundaries of the state space, the dynamical system is defined as follows:

$$
\nu_1^k(t) = 0 \Rightarrow \dot{\nu}_1^k(t) = \max\{0, \varphi^k(\nu(t))\}
\tag{4.23}
$$
$$
\nu_1^k(t) = 1 \Rightarrow \dot{\nu}_1^k(t) = \min\{0, \varphi^k(\nu(t))\}
\tag{4.24}
$$

for each $k = 1, 2$.

One may provide a variety of different interpretations for the dynamics described by (4.21)–(4.24). One possibility is in terms of a generally conceived process of "migration". Just to make matters vivid and simple, suppose again that each of the two populations lives in its corresponding island and, in each of these islands, there are separate geographical areas where individuals of each type live. When news arrives that the average pay-off in one part of the island

---

[14] Of course, only the dynamics for one of the components of each population's profile needs to be explicitly specified (in this case, it is the first component for each one of them).

is higher than in the other, migration from the former to the latter starts to take place. The rates $\alpha$ and $\beta$ at which this migration occurs should be related, for example, to the characteristics of the island (orography, size, etc.). In any case, these rates are postulated to be identical for migration in both directions and independent of the magnitude of the pay-off gap.

We now turn to the analysis and comparison of the two evolutionary systems proposed. For the RD (4.19)–(4.20), first observe that Table 11 can be transformed into that of a zero-sum game through manipulations which do not affect the dynamics. For example, take the pay-off matrix of the second population and simply add 2 to its first-row pay-offs and $-1$ to those in the second row. Analogously, take the pay-off matrix of the first population and add $-4$ to the first-column entries and $-7$ to those of the second column. After performing these operations, the following pay-off table results:

| 2<br>1 | $s_1^2$ | $s_2^2$ |
|:---:|:---:|:---:|
| $s_1^1$ | $-3, 3$ | $0, 0$ |
| $s_2^1$ | $0, 0$ | $-3, 3$ |

Table 12

Denote by $A$ and $B$ the resulting pay-off matrices of players 1 and 2. Then, $A = -B$. Moreover, the game has the (unique) interior Nash equilibrium $((\frac{1}{2}, \frac{1}{2}), (\frac{1}{2}, \frac{1}{2}))$. Thus, along the lines of Theorem 14 it may be asserted that the function defined by

$$\zeta(\nu) \equiv \log \nu_1^1 + \log(1 - \nu_1^1) + \log \nu_1^2 + \log(1 - \nu_1^2)$$

represents a *constant of motion* for the evolutionary system. As established by this Theorem (which may readily be verified directly), $\dot{\zeta}(\nu(\cdot)) \equiv 0$. This indicates that, as illustrated in Figure 6, the system displays closed counter-clockwise orbits around the point $(\frac{1}{2}, \frac{1}{2})$ in the $(\nu_1^1, \nu_1^2)$-space.

On the other hand, it is easy to see that the orbits of the dynamical system (4.21)–(4.24) are closed, counter-clockwise, and piece-wise linear around $(\frac{1}{2}, \frac{1}{2})$, with an absolute slope equal to $\beta/\alpha$ in the interior of the state space. If a trajectory enters the boundary at some point, it coincides with it up to the point where it can "merge" with a trajectory which is reflected by the boundary and enters the interior of the state space. An illustration is contained in Figure 7.

Thus, qualitatively, the behaviour of the two dynamics considered is quite similar. Essentially, they both yield closed orbits around their unique (interior) rest point $(1/2, 1/2)$. They only differ in one important point, already discussed in Section 4.6 when contrasting (4.1) with the more general formulation (4.14).

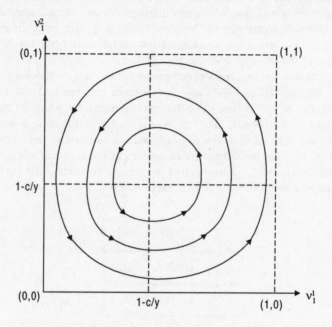

Figure 6: Risky trading, Replicator Dynamics

Whereas the Replicator Dynamics (4.19)–(4.20) never allows an extinct strategy to appear (or, for that matter, a previously present strategy to go extinct), the gradient-monotonic system (4.21)–(4.24) permits both of these possibilities.

It seems natural to require that an evolutionary system should display the potential of escaping the boundary of the state space. Gradient monotonicity provides an interesting possibility in this respect. Another possibility, the explicit modelling of mutation, will be one of the main themes of the next two chapters. Still a third approach, the perturbation of the dynamical system so that a permanent flow of individuals adopt (mutate to) every given strategy, has already been illustrated in Section 3.9. An additional, similar context is presented in the next section, within the framework of a simple two-population bargaining game.

## 4.8    A Simplified Ultimatum Game

Consider two populations, 1 and 2, whose members are randomly matched to play an ultimatum game. This game involves dividing a given total surplus

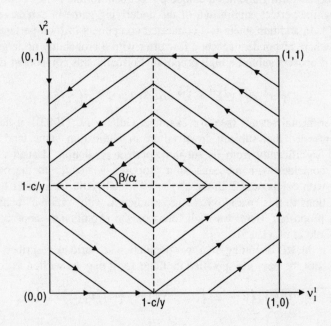

Figure 7: Risky trading, a gradient-monotonic evolutionary system

between the two players matched, with the player of population 1 (the proposer) making a take-it-or-leave-it offer and the player of population 2 (the responder) either accepting or rejecting it. If the offer is rejected, the surplus is lost to both players.

Following Gale *et al.* (1995), a simplified and concrete version of this game will be analysed in which the total surplus to be divided is 4, and the proposer just makes one of two proposals: a high one equal to 2 (strategy $s_1^1$) or a low one equal to 1 (strategy $s_2^1$). Assuming a responder always accepts a high offer, this game may be summarized by the following pay-off table (where $s_1^2$ stands for "acceptance" and $s_2^2$ for "rejection" of a low offer):

| 1 \ 2 | $s_1^2$ | $s_2^2$ |
|---|---|---|
| $s_1^1$ | 2, 2 | 2, 2 |
| $s_2^1$ | 3, 1 | 0, 0 |

Table 13

This normal-form game has a unique perfect equilibrium which corresponds to the subgame-perfect equilibrium of the underlying game in extensive form, i.e. $(s_2^1, s_1^2)$. In addition, there is a connected component of (non-perfect) Nash equilibria where responders accept a low offer with a probability no larger than $\frac{2}{3}$ and the proposers submit a high proposal. Formally, this component is given by

$$\Gamma = \{(\sigma^1, \sigma^2) \in \Sigma^1 \times \Sigma^2 : \sigma_1^1 = 1, \ 0 \le \sigma_1^2 \le \tfrac{2}{3}\}.$$

In experimental set-ups (see, for example, Güth *et al.* (1982)), it has been widely confirmed that subjects involved in an Ultimatum game tend to deviate quite significantly from its subgame-perfect prediction. Usually, modal behaviour coincides with the equal-split proposal, analogous to the outcome associated with component $\Gamma$ in the above-described game. A number of alternative solutions to this "paradox" (i.e. contradiction with "rational" behaviour) have been proposed. Here, we shall focus on the evolutionary approach proposed by Gale *et al.* (1985).

Consider the RD operating on a two-population scenario as described above. On the basis of the pay-offs specified in Table 13, it may be written as follows:

$$\dot{\nu}_1^1(t) \;=\; \nu_1^1(t)\,\left\{2 - \left[2\nu_1^1(t) + 3\,\left(1 - \nu_1^1(t)\right)\nu_1^2(t)\right]\right\} \tag{4.25}$$

$$\dot{\nu}_1^2(t) \;=\; \nu_1^2(t)\,\{2\nu_1^1(t) + \left(1 - \nu_1^1(t)\right) - \tag{4.26}$$
$$\left[\nu_1^2(t)\left(2\nu_1^1(t) + \left(1 - \nu_1^1(t)\right)\right) + 2\,\left(1 - \nu_1^2(t)\right)\nu_1^1(t)\right]\}.$$

The dynamics induced by (4.25)–(4.26) may be seen to display the qualitative features shown in Figure 8. On the one hand, only the population state $\nu^* = ((0,1),(1,0))$ qualifies as asymptotically stable, having a neighbourhood around it where all trajectories converge to this point. On the other hand, all those trajectories which do not fall into the basin of attraction of $\nu^*$ must converge to the set $\Gamma$.[15] In this set, every state is Liapunov-stable (i.e. satisfies condition (i) of Definition 6 but *not* its condition (ii)), with the sole exception of its boundary point $((1,0),(\frac{2}{3},\frac{1}{3}))$.

The qualitative behaviour just described is quite reminiscent of that described in Section 3.9 for a single-population context involving the repeated prisoner's dilemma. Thus, in analogy with the approach undertaken there, it is natural to ask whether a slight perturbation of the dynamics (4.25)–(4.26) may have significant qualitative effects on its long-run behaviour. In particular, the key issue centres on analysing the behaviour of the perturbed system close to the component $\Gamma$, where the rest points of the system could not be structurally stable. (Elsewhere, the introduction of just *small* noise cannot have any qualitative effect on the dynamics of the system.)

---

[15] Of course, this statement only applies to those interior trajectories which do not start at a simplex vertex for *both* populations. Also note that, for notational simplicity, the set of population states and the set of mixed-strategy profiles are both denoted by $\Gamma$.

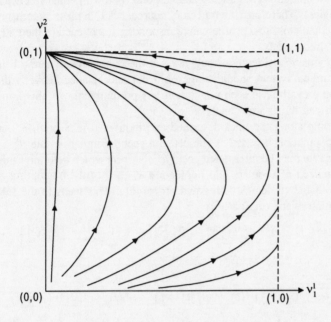

Figure 8: Simplified ultimatum game, unperturbed dynamics

A priori, two alternative types of behaviour might be expected around this component from the introduction of small "mutational noise" into the system:

(a) No rest points remain close to $\Gamma$, the composition of selection and mutational forces leading every trajectory of the system towards state $\nu^*$.

(b) The whole component $\Gamma$ shrinks to a *single* point close to it which is asymptotically stable for the perturbed dynamics.

Alternative (a) would clearly prevail if mutation were to operate *only* on proposers. For, in this case, the fact that there would always be a small (but positive) fraction of proposers adopting strategy $s_2^1$ implies that the expected pay-off for $s_1^2$ would be higher than for $s_2^2$. Consequently, selection forces (the only ones at work on responders in this case) would induce $\dot{\nu}_1^2(t) > 0$, eventually making the system exit any small neighbourhood of $\Gamma$, subsequently converging towards $\nu^*$.

In contrast, alternative (b) would materialize if mutation were to impinge *only* on responders. In this case, selection pressures on the latter operating in a sufficiently small neighbourhood of $\Gamma$ would be so small that mutation would play the decisive dynamical role. But within $\Gamma$, mutation, by itself, determines a unique asymptotically stable point (one with equal frequencies for both strategies). Therefore, if a trajectory approaches $\Gamma$ in these circumstances, it must also come close to a profile where responders adopt each of their strategies with equal frequencies.

The previous considerations suggest that the qualitative dynamic behaviour of the perturbed system should hinge on a delicate balance between the rates of mutation prevailing in each population. As presently shown, this intuition is fully confirmed by rigorous analysis.

Proceeding along the lines described and motivated in Section 3.9, assume that each population $k = 1, 2$ is subject to a specific mutation rate $\theta^k > 0$. As before, this rate may be interpreted as a *turnover rate* which determines the flow of inexperienced newcomers into population $k$, each of them adopting each of the two available strategies with equal probability. This leads to the following transformation of the original RD:

$$\dot{\nu}_1^1(t) = (1 - \theta^1)\, \nu_1^1(t)\, \left\{2 - \left[2\nu_1^1(t) + 3\left(1 - \nu_1^1(t)\right)\nu_1^2(t)\right]\right\}$$
$$+\, \theta^1 \left(\frac{1}{2} - \nu_1^1(t)\right)$$

$$\dot{\nu}_1^2(t) = (1 - \theta^2)\, \nu_1^2(t)\, \left\{2\nu_1^1(t) + \left(1 - \nu_1^1(t)\right) - \left[\nu_1^2(t)\left(2\nu_1^1(t)\right.\right.\right.$$
$$\left.\left.\left. +\left(1 - \nu_1^1(t)\right)\right) + 2\left(1 - \nu_1^2(t)\right)\nu_1^1(t)\right]\right\} + \theta^2 \left(\frac{1}{2} - \nu_1^2(t)\right).$$

which can be simplified as follows:

$$\dot{\nu}_1^1(t) = (1 - \theta^1)\, \nu_1^1(t)\, \left(1 - \nu_1^1(t)\right)\left(2 - 3\nu_1^2(t)\right) + \theta^1 \left(\frac{1}{2} - \nu_1^1(t)\right) \tag{4.27}$$

$$\dot{\nu}_1^2(t) = (1 - \theta^2)\, \nu_1^2(t)\, \left(1 - \nu_1^2(t)\right)\left(1 - \nu_1^1(t)\right) + \theta^2 \left(\frac{1}{2} - \nu_1^2(t)\right) \tag{4.28}$$

Our analysis will be concerned with the dynamics induced by (4.27)–(4.28) when the mutation rates are very small. Formally, this involves studying the behaviour of this dynamical system in the limit, when both $\theta^1$ and $\theta^2$ converge to zero. As suggested above, the relative speeds at which these parameters become small should play a crucial role in the outcome of this limit exercise. As it turns out, if given any $\phi > 0$, we fix

$$\phi = \frac{(1 - \theta^1)\, \theta^2}{(1 - \theta^2)\, \theta^1} \tag{4.29}$$

as $\theta^1 \to 0$ and $\theta^2 \to 0$, two different cases need to be considered.

*Case 1:*  $\phi < 3 + 2\sqrt{2}$.

*Case 2:*  $\phi > 3 + 2\sqrt{2}$.

Cases 1 and 2 define alternative scenarios where, respectively, the proposer or responder population is subject to a relatively higher mutation rate. Thus, whereas Case 1 may be viewed as approaching the extreme context where only proposers are subject to mutation, Case 2 may be seen as an approximation to the extreme polar context where only responders are subject to it. (Recall the previous discussion concerning (a) and (b) above.) For each of these two cases, the next result establishes limit behaviours for the system which are qualitatively very different.

**Proposition 10 (Gale** *et al.* **(1995))** *Denote by* $\mathcal{A}(\theta^1, \theta^2)$ *the set of asymptotically stable states of the system for mutation (or turnover) rates* $(\theta^1, \theta^2)$, *and let* $\mathcal{A}^* \equiv \lim_{\theta^1, \theta^2 \to 0} \mathcal{A}(\theta^1, \theta^2)$ *with* $\frac{(1-\theta^1)\theta^2}{(1-\theta^2)\theta^1} = \phi$ *for some fixed* $\phi$.
*(i) In Case 1,* $\mathcal{A}^* = \{\nu^*\}$.
*(ii) In Case 2,* $\mathcal{A}^* = \{\nu^*, \hat{\nu}\}$, *with* $\hat{\nu}_1^1 = 1$ *and* $\hat{\nu}_1^2 < \frac{2}{3}$.

**Proof.** First, the stationary points of the system are computed by making $\dot{\nu}_1^1(t) = \dot{\nu}_1^2(t) = 0$ in (4.28)–(4.29). This yields the following equations

$$-\theta^1 \left(\frac{1}{2} - \nu_1^1\right) = (1-\theta^1) \, \nu_1^1 \, (1 - \nu_1^1) \, (2 - 3\nu_1^2) \qquad (4.30)$$

$$-\theta^2 \left(\frac{1}{2} - \nu_1^2\right) = (1-\theta^2) \, \nu_1^2 \, (1 - \nu_1^2) \, (1 - \nu_1^1). \qquad (4.31)$$

Dividing (4.31) by (4.30), and after some rearrangement, one obtains

$$\phi = \frac{(1-\theta^1)\,\theta^2}{(1-\theta^2)\,\theta^1} = \frac{\nu_1^2(\frac{1}{2} - \nu_1^2)(1 - \nu_1^1)(1 - \nu_1^2)}{\nu_1^1(\frac{1}{2} - \nu_1^2)(1 - \nu_1^1)(2 - 3\nu_1^2)}, \qquad (4.32)$$

where the term $(1 - \nu_1^1)$ in the numerator and denominator of the above expression can be cancelled since every rest point of the system must have $\nu_1^1 < 1$. Looking for solutions close to $\Gamma$, we make $\nu_1^1 \to 1$ in (4.32) to obtain the equation

$$\phi = \frac{-\nu_1^2(1 - \nu_1^2)}{(1 - 2\nu_1^2)(2 - 3\nu_1^2)}.$$

When $\phi < 3 + 2\sqrt{2}$ (Case 1), it can be checked that the above equation has no well-defined solutions in the interval $[0, 1]$. In alternative Case 2, it has two solutions, $\hat{\nu}_1^2$ and $\tilde{\nu}_1^2$, satisfying:

$$\frac{1}{2} < \hat{\nu}_1^2 < 2 - \sqrt{2} < \tilde{\nu}_1^2 < \frac{2}{3}. \qquad (4.33)$$

To complete the proof, it is enough to show that, for small $\theta^1$ and $\theta^2$ which satisfy (4.29), the rest point close to $\hat{\nu} = ((1, 0), (\hat{\nu}_1^2, 1 - \hat{\nu}_1^2))$ is stable whereas the one near $\tilde{\nu} = ((1, 0), (\tilde{\nu}_1^2, 1 - \tilde{\nu}_1^2))$ is not. To assess this stability, let $J(\nu_1^1, \nu_1^2)$ stand for the Jacobian-matrix function associated with the vector field defining (4.27)–(4.28). This $2 \times 2$-matrix is of the following form:

$$\begin{pmatrix} (1-\theta^1)(2 - 3\nu_1^2)(1 - 2\nu_1^1) - \theta^1 & -3(1-\theta^1)\,\nu_1^1(1 - \nu_1^1) \\ -(1-\theta^2)\,\nu_1^2(1 - \nu_1^2) & (1-\theta^2)\,(1 - 2\nu_1^2)(1 - \nu_1^1) - \theta^2 \end{pmatrix}. \tag{4.34}$$

It is enough to show that the eigenvalues of $J(\hat{\nu}_1^1, \hat{\nu}_1^2)$ have both negative real parts whereas at least one of $J(\tilde{\nu}_1^1, \tilde{\nu}_1^2)$ has a positive real part. From (4.33), it follows that the trace of both of these matrices is negative. Thus, the claim hinges upon their respective determinants. To evaluate them, multiply the second column of (4.34) by $2\nu_1^2 - 1$ (which, by (4.33), is positive at every rest point) and use the equality

$$\theta^2(2\nu_1^2 - 1) = 2(1 - \theta^2)\,\nu_1^2(1 - \nu_1^2)(1 - \nu_1^1)$$

which results from (4.31) and, therefore, is satisfied at every rest point. Factoring out $(1 - \theta^1)(1 - \theta^2)(1 - \nu_1^1) > 0$ and making $\theta^1 \to 0$ and $\nu_1^1 \to 1$, the sign of the determinant $\left| J(\nu_1^1, \nu_1^2) \right|$ is that of

$$\begin{vmatrix} 2 - 3\nu_1^2 & 3(2\nu_1^2 - 1) \\ \nu_1^2(1 - \nu_1^2) & 1 - 2\nu_1^2(1 - \nu_1^2) \end{vmatrix} = \left[\nu_1^2\right]^2 - 4\nu_1^2 + 2.$$

Since the roots of the previous polynomial are $2 \pm \sqrt{2}$, (4.33) confirms the desired conclusion, completing the proof of the proposition. ∎

The previous result establishes that, in cases where the noise which affects responders is sufficiently *higher* than that on proposers, a unique (but non-perfect) Nash equilibrium, close to the component $\Gamma$, will be asymptotically stable (cf. Figure 9). Thus, in this case, evolutionary arguments are consistent with the existence of a dynamically robust equilibrium which, in line with empirical evidence, does not necessarily assign all (or most of) the surplus to the proposer.

The previous considerations raise the natural question of what might underlie any interpopulation differences in mutation rates. Adopting the interpretation formerly proposed, i.e. that these rates reflect a flow of newcomers entering the game, Gale *et al.* (1995) suggest the following motivation. Any new individual who comes into the game will typically be involved in many such games. Therefore, it is reasonable to postulate that the attention he will devote to analysing any particular new game should depend on what are the pay-offs at stake. If these are small, he will tend to decide on his action in a basically random fashion, selecting any of the available strategies with equal probability. On the contrary, if the stakes are relatively large, his choice will be more closely tailored to the characteristics of the game and, on average, closer to the currently best action. Within the contemplated formulation, this would amount to saying that the rate $\theta^k$ at which "heedless" players flow into population $k$ will be larger the smaller are the pay-offs at stake.

Within the Ultimatum game under consideration, the previous discussion provides some heuristic basis for the idea that, as a trajectory close to the component $\Gamma$ sees the stakes for proposers rise, the (purely unbiased) noise which affects this population should decrease (reciprocally, that on responders should increase). Of course, a complete formalization of this idea would require

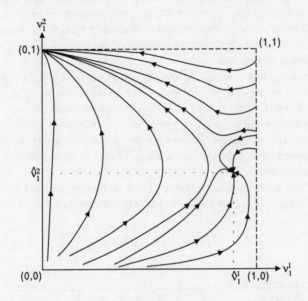

Figure 9: Simplified ultimatum game, noisy dynamics

introducing into the model a variable mutation rate whose magnitude is linked to the current state of the system. This, however, would complicate the model substantially, rendering a formal analysis of it significantly more difficult.[16]

## 4.9 A Hierarchic Model of Cultural Evolution

As suggested in Section 3.10, some evolutionary processes might be suitably conceived as proceeding at different and interacting levels. It was also argued there that, in order for such a multi-level approach to be fruitful, the evolutionary process under consideration must proceed at a relatively fast pace at every one of the considered levels. Only then would these different levels genuinely interact, thus rendering it worth while to study a simultaneous dynamical system that encompasses all of them.

Whereas in biology the validity of such an implicit assumption (i.e. relatively fast evolution at different levels) is typically dubious, it may nevertheless

---

[16] Even though Gale *et al.* (1995) confirm the outlined intuition through extensive simulations, they rely on the condition specified for Case 1 above in order to address it formally in a more tractable manner.

represent an appropriate working hypothesis for modelling certain economic and other social environments. Here, no attempt will be made to provide a general framework for the study of hierarchical evolutionary models in these contexts. Rather, we shall simply illustrate the nature of this approach by means of a simple economic application.[17]

Consider a dynamic context where at each point in (continuous) time $t \geq 0$ there is a large population of firms. Specifically, assume that there is a continuum of them, with a total and fixed measure equal to one. Each of these firms employs the same (finite) number of workers, say $m$, its production activities modelled as a simple $m$-person game played among its workers. This game is supposed to include just two actions, $H$ and $L$, interpreted as high- and low-effort levels, respectively.

Pay-offs are postulated as follows. Let $e = (e_1, e_2, ..., e_m)$ be the effort profile prevailing in some given firm. The pay-off earned by worker $i$ is given by

$$\pi^i(e) = x(e) - \frac{1}{2}\delta(e_i),$$

where

$$x(e) = 1 \quad \text{if } e_i = H \text{ for all } i = 1, 2, ..., m,$$
$$= 0, \quad \text{otherwise,}$$

and

$$\delta(H) = 1; \quad \delta(L) = 0.$$

Clearly, this game has two (strict) Nash equilibria, $e^0 = (L, L, ..., L)$ and $e^1 = (H, H, ..., H)$, where workers co-ordinate in one of the effort levels, low or high, respectively.

The proposed game is a rather extreme representative of the so-called stag-hunt variety.[18] In particular, this implies that, within each firm, only the Pareto-inferior (or inefficient) equilibrium $e^0$ is *evolutionarily stable*. (Recall the finite-population version of this concept introduced in Definition 3.) In fact, it is clear that if one postulates any evolutionary dynamics operating *within* the firm which satisfies one of the (suitably adapted) monotonicity concepts proposed above, the basin of attraction of $e^0$ includes *all* profiles $e \neq e^1$.[19] This suggests assuming that, at the lower (intrafirm) level, evolution will always lead to the inefficient

---

[17] This application relies heavily upon Vega-Redondo (1993). Other instances of multi-level evolutionary analysis can be found in Robson (1993), Oechssler (1993), or Vega-Redondo (1996a).

[18] There is nothing essential in the extreme nature of the stag-hunt game considered. As shown in Vega-Redondo (1993), the analysis can be extended to analogous stag-hunt games where the threshold for $L$ to be the relatively better strategy is not just one player but any number smaller than $m$.

[19] As emphasized by Crawford (1991), it is precisely this strong property of $e^0$ which seems to underlie the rather clear-cut experimental results obtained by van Huyck *et al.* (1990) for this type of games. These authors have found that, in a large fraction of cases, experimental subjects who play these games repeatedly end up co-ordinated in the inefficient equilibrium.

equilibrium whenever the firm in question has *some* player choosing a low-effort level. For future reference, this is contained in the following postulate:

*Postulate 1:* If a firm is playing one of the two equilibria, it remains playing it. Otherwise, it evolves towards $e^0$.

This addresses the lower level dynamics of the process. For simplicity, it will be assumed that the final outcome of this intra-firm process (either $e^0$ or $e^1$) materializes before any of the inter-firm considerations described below are to apply.

The upper level (inter-firm) dynamics is taken to include two different components. The first one pertains to the (differential) rate at which alternative types of firms survive. A natural "monotonicity condition" in this respect would be to link the chances of survival of each firm to its relative total pay-off. Since, given Postulate 1, firms may be partitioned into two groups (those that play the efficient equilibrium – the "efficient firms" – and those that play the inefficient one, i.e. the "inefficient firms"), a very stylized way of capturing this idea may be formulated as follows.

*Postulate 2:* Efficient and inefficient firms dissolve at respective rates $p$, $q$; $p < q$.

The former postulate captures only one of the dimensions of interfirm selection: inefficient firms disappear at a faster rate than efficient ones. The other side of the coin involves the rate at which new firms are created. Different possibilities may be contemplated in this respect. To discuss them formally, some notation is now introduced.

Let $\mu(t)$ denote the fraction of efficient firms prevailing at $t$. By Postulate 2, these efficient firms dissolve at the rate $p\mu(t)$. Correspondingly, inefficient firms do so at the rate $q(1 - \mu(t))$. Since the total measure of firms is assumed to remain fixed, the rate at which new firms are created must be $d(t) \equiv p\mu(t) + q(1 - \mu(t))$.

From Postulate 1, one just needs to consider the rates at which efficient and inefficient firms are eventually formed (i.e. once the intrafirm dynamics is complete). Thus, let $\chi(\mu(t))$ be the proportion of firms created at $t$ which become efficient. If one adopted the view that firms expand (or "reproduce") by generating entities similar to themselves (i.e. just as biological individuals would do), a counterpart of the monotonicity contemplated by Postulate 2 could be stated as follows.

*Postulate 3:* For all $\mu \in [0, 1]$, $\chi(\mu) \geq \mu$.

However, this formulation is highly questionable on the grounds that, in fact, groups are *not* individuals. Therefore, they will generally find it very hard, if not impossible, to replicate themselves, achieving the fine-tuned degree of co-ordination required for an efficient equilibrium so fragile as the one considered. Only to the extent that some "firm culture" can be passed on quite rigidly,

Postulate 3 could be a reasonable theoretical approximation of the underlying evolutionary forces. In this case, it is easy to see that the pressure towards efficiency must eventually prevail. Simply write the (one-dimensional) upper-level dynamics as follows:

$$\dot{\mu}(t) = -p\,\mu(t) + \chi(\mu(t))\,d(t). \tag{4.35}$$

Since $d(t)$ is a convex combination of $p$ and $q$ which remains strictly larger than $p$ if $\mu$ is bounded below 1, Postulate 3 implies that $\mu(t) \to 1$ as $t \to \infty$.

To obtain a more interesting model, Postulate 3 must be replaced by an inheritance mechanism which is much less group-based. For example, in the spirit of Boyd and Richerson (1985), it may be assumed that those workers who enter new firms adopt their *initial* strategy by global mechanisms of cultural inheritance. As a simple embodiment of this idea, let us assume that workers enter new firms with an initial propensity (i.e. *ex ante* probability) for high effort equal to the frequency with which $H$ is adopted in the overall population. In view of Postulate 1, the effect of this cultural mechanism on the rate of creation of efficient firms may be compactly described as follows.

*Postulate 3':*   For all $\mu \in [0,1]$, $\chi(\mu) = [\mu(t)]^m$.

As established by the following proposition, the dynamic analysis resulting from Postulates 1, 2, and 3' is now substantially richer than before. In particular, the long-run behaviour of the system hinges upon a clear-cut comparison between the "efficiency bite" $\alpha \equiv \frac{q}{p}$ and the size $m$ of the firms (an intuitive measure of how difficult it is to achieve efficient co-ordination).

**Proposition 11 (Vega-Redondo (1993))** *Under Postulates 1, 2, and 3', the long-run dynamics of the evolutionary process may be characterized as follows:*
*(a) If $\alpha \le m$, $\lim_{t\to\infty} \mu(t) = 0$, for all initial $\mu(0) < 1$.*
*(b) If $\alpha > m$, there exists some stationary $\mu^* \in (0,1)$ such that:*
    *(b.1) $\lim_{t\to\infty} \mu(t) = 0$ if $\mu(0) < \mu^*$;*
    *(b.2) $\lim_{t\to\infty} \mu(t) = 1$ if $\mu(0) > \mu^*$.*

**Proof.** Define the function $f : [0,1] \to \Re$ by

$$f(\mu) = (p - q)\mu^{m+1} + q\,\mu^m - p\,\mu. \tag{4.36}$$

From (4.35) and Postulate 3', the dynamics of the process can be written as

$$\begin{aligned} \dot{\mu}(t) &= -p\,\mu(t) + [\mu(t)]^m\,[p\,\mu(t) + q\,(1 - \mu(t))] \\ &= (p - q)\,[\mu(t)]^{m+1} + q\,[\mu(t)]^m - p\,\mu(t). \end{aligned}$$

Therefore, stationary points of the dynamics are zeros of the function $f(\cdot)$. In particular, $f(0) = f(1) = 0$, which implies that the two monomorphic configurations are stationary.

The desired characterization then easily follows once the following three claims are verified.

*Claim 1:*  For all $\alpha > 1$, $m \in \mathcal{N}$, $f'(0) \equiv \frac{df}{d\mu}(0) < 0$.

*Claim 2:*  If $\alpha > m$, $f'(1) < 0$. Otherwise (i.e. if $\alpha \leq m$), the function $f(\cdot)$ is locally increasing at $\mu = 1$.

*Claim 3:*  There is at most one $\mu^* \in (0,1)$ such that $f(\mu^*) = 0$. Moreover, if such $\mu^*$ exists, $\frac{df}{d\mu}(\mu^*) \neq 0$.

The proofs of Claim 1 and the first part of Claim 2 follow from direct computation. This is also the case for the second part of Claim 2 if $\alpha < m$. In the boundary case $\alpha = m$, we have $f'(1) = 0$ and, therefore one needs to resort to straightforward second-order considerations.

To prove Claim 3, define the function $g : [0,1] \to \Re$ by

$$g(\mu) = (p - q)\,\mu^m + q\,\mu^{m-1} - p.$$

Since, for every $\mu > 0$,

$$g(\mu) = \frac{1}{\mu}\,f(\mu)$$

it is clear that the interior zeros of $f(\cdot)$ are also zeros of $g(\cdot)$. Now compute:

$$\frac{dg}{d\mu}(\mu) \equiv g'(\mu) = (p - q)\,m\,\mu^{m-1} + q\,(m-1)\,\mu^{m-2}$$

and define $h : (0,1) \to \Re$ by

$$h(\mu) = \frac{1}{\mu^{m-2}}\,g'(\mu) = (p - q)\,m\,\mu + q(m - 1).$$

The function $h(\cdot)$ is linear in $\mu$ with negative slope equal to $-m(q-p)$. Moreover, one has

$$sgn\,[g'(\mu)] = sgn\,[h(\mu)],$$

for all $\mu \in (0,1)$. Therefore, there can only be at most one interior point $\hat{\mu}$ where $g'(\hat{\mu}) = 0$. Since

$$g(0) = -p < 0 = g(1),$$

this implies the desired conclusion, i.e. there exists at most one $\mu^* \in (0,1)$ such that $f(\mu^*) = 0$ and, if such $\mu^*$ exists, $\frac{df}{d\mu}(\mu^*) \neq 0$.

Combining Claims 1 to 3, the desired characterization follows immediately, thus completing the proof of the Theorem.  ∎

The preceding result provides an intuitive characterization of the conditions under which the efficient social convention (high effort) will be locally stable, thus having at least *some* basin of attraction with positive measure. Essentially, what is required is that the forces of selection (or "competition") be strong enough relative to the difficulty (or "fragility") of efficient co-ordination. Indeed, it is easy to derive parameter-sensitivity results which reflect this idea in a stark fashion. For example, it can be shown that the dividing frontier $\mu^*$ between the two basins of attraction moves arbitrarily close to the lower (or upper) end of the state space $[0, 1]$ as the parameter $\alpha$ reflecting the severity of competition falls (respectively, grows) or as firm size $m$ grows (respectively, falls).

# 5
# Stochastic Evolution

## 5.1 Introduction

Out of the three basic evolutionary forces described in Chapter 1 – selection, inheritance, and mutation – the latter has only played an implicit role in most of our developments so far. (For example, even though an idea of infrequent mutation underlay the concept of ESS presented in Definition 1, this remained at a purely implicit level.) Only Sections 3.9 and 4.8 have discussed some illustrative models where the force of mutation is an explicit component of the analysis. There, the reader will recall, it amounts to a perturbation of the selection-based dynamic system whose effect is to introduce a *deterministic* flow of individuals adopting every possible strategy.

Within the large (formally infinite) population context considered in those examples, this deterministic approach to the phenomenon of mutation may be conceived as a suitable theoretical idealization. As discussed, it sheds new light on the relative robustness of alternative outcomes of the model. It may be argued, however, that mutation is an intrinsically stochastic phenomenon whose appropriate formulation requires a fully stochastic (as well as dynamic) model. It is essentially this view which has led to the large body of recent literature whose models are explicitly stochastic and dynamic. In most of these models, the population is assumed finite and therefore subject to the a priori uncertain effects of random mutation. Hence, instead of deterministic trajectories, only the prior probabilities over alternative paths may be determined *ex ante*.

If the evolutionary process is ergodic, one may also be able to specify the long-run (average) behaviour of the system, somewhat in the spirit of the analysis conducted in Subsection 3.7.4. As it turns out, this long-run analysis will sometimes produce a clear-cut selection outcome (independent of initial conditions) in cases where traditional Game Theory typically remains silent (for example, in strict co-ordination games). In other cases, however (see the next introduce "volatility" where, a priori, some definite selection criteria would seem quite forceful.

## 5.2   A Simple Example

Consider a certain finite population of $n$ individuals ($n$ even) which are matched to play the following (symmetric) co-ordination game:[1]

|       | $s_1$            | $s_2$ |
|-------|------------------|-------|
| $s_1$ | $\sqrt{3}, \sqrt{3}$ | $0, 1$ |
| $s_2$ | $1, 0$           | $1, 1$ |

Table 14

Note that, in this game, strategy $s_1$ is the efficient (Nash-equilibrium) strategy. However, strategy $s_2$ also defines a symmetric Nash equilibrium, with the additional feature that it is risk-dominant in the sense of Harsanyi and Selten (1988). Specifically, $s_2$ is the optimal strategy for an individual who (under the implicit assumption that he has no information on the basis of which to discriminate between the two strategies) attributes the same probability to his opponent playing either of them.

Modelling time in a discrete fashion,[2] assume that every period $t = 1, 2, ...$ there is some given number of independent matching rounds, and denote by $z_t \in \{0, 1, ..., n\}$ the number of individuals who play strategy $s_1$ in *each* one of them (of course, $n - z_t$ play strategy $s_2$). The general idea embodied by the evolutionary system proposed is twofold:

(a)  Changes in the frequency with which each strategy is played at $t + 1$ must "monotonically" depend on the relative average pay-offs prevailing at $t$.

(b)  Occasionally (i.e. with some small probability) players "mutate" (or experiment), their new strategy choices being then unrelated to the strategies' current pay-offs.

The first point captures the *selection* (core) component of the evolutionary dynamics. It reflects the generalized Darwinian idea that the fittest strategies

---

[1] This example is borrowed from Robson and Vega-Redondo (1996).

[2] Most of the recent evolutionary literature in a stochastic framework considers a discrete-time set-up. Some exceptions e.g. Foster and Young (1990), Fudenberg and Harris (1992), are summarized in Section 5.5.

(i.e. those with the highest average pay-off) should see their population frequency increase. The second point responds to the idea that players will also sporadically experiment. In a stylized fashion, this is identified with the notion that, with small probability, agents will choose their strategy in a way fully unrelated to current circumstances.

Heuristically, one may think of an evolutionary system as the composition of "selection plus noise". As explained in (b) above, one source for such noise is mutational (or experimental). However, in finite populations, another source of potentially crucial noise is induced by the fact that the matching mechanism is assumed random.[3] In some contexts, this latter type of noise, in rich interaction with mutational noise, may have substantial effects on the outcome of the model. To illustrate these matters, two alternative scenarios will be compared here: one where the matching-based noise is high relative to mutational noise; a second one, where the opposite applies. For the illustrative purposes of our present example, the comparison will be established between the following two extreme contexts:

- Scenario 1, where there is just *one* round of matching per period.

- Scenario 2, where there are an "infinite" number of independent matching rounds per period.

Consider first Scenario 1. Because of random matching, the average pay-offs earned by each strategy will depend not only on the current number of individuals $(z_t, n - z_t)$ who play each strategy but also on the number of cross-matchings realized, i.e. the number of pairings where each of the two individuals adopts a *different* strategy.

Let $\tilde{p}_t$ stand for the *random* variable determining the number of cross-matchings realized at $t$, and denote by $p_t$ a typical realization of it. If $z_t$ is even, it is clear that the set of possible realizations is given by the set $\{0, 2, ..., \min[z_t, n - z_t]\}$. On the other hand, if $z_t$ is an odd number, the support of $\tilde{p}_t$ is the set $\{1, 3, ..., \min[z_t, n - z_t]\}$.

Let $\pi_i(z_t, p_t)$ denote the average pay-off earned by strategy $s_i$ when the state of the system is $z_t$ and the number of realized cross-matchings is $p_t$. On the basis of Table 14, and provided $0 < z_t < n$, they may be computed as follows:

$$\pi_1(z_t, p_t) = \frac{\sqrt{3}(z_t - p_t)}{z_t};$$
$$\pi_2(z_t, p_t) = 1.$$

Given any such realized average pay-offs, denote by $B(z_t, p_t)$ the number of individuals who choose to adopt strategy $s_1$ at $t + 1$ due to selection forces, i.e. by the considerations explained in (a) above. If, to simplify matters, it

---

[3] In a finitely repeated prisoner's dilemma, Young and Foster (1991) still consider a third kind of noise which they show to have interesting long-run implications, namely the number of periods during which interaction takes place between any two given players.

is assumed that *all* players react immediately to any pay-off differences, the selection component of the dynamics may be formulated as follows:

$$B(z_t, p_t) = \begin{cases} n & \text{if} \quad \pi_1(z_t, p_t) > \pi_2(z_t, p_t) \\ 0 & \text{if} \quad \pi_1(z_t, p_t) < \pi_2(z_t, p_t) \end{cases} \tag{5.1}$$

when $0 < z_t < n$. (Note that the pay-off of $\sqrt{3}$ in Table 14 – an irrational number – implies that pay-off ties between the two strategies can never occur.)

If either $z_t = 0$ or $z_t = n$, the system is assumed to remain fixed, i.e. $B(0,0) = 0$ or $B(n,0) = n$, respectively. The implicit idea here is that no *new* strategies may arise except by mutation (recall (b) above). To model mutation the following simple formulation is proposed:

> Every period $t$, each player switches to a new strategy with some independent probability $\epsilon > 0$. In that event, both strategies are chosen with positive probability, independently of what (5.1) prescribes.

In order to keep track of the evolution of the process in a convenient fashion, identify the state of the system at any $t$ with the number of individuals who adopt strategy $s_1$ right *after* selection has taken place, but *before* any individual may mutate. Clearly, from (5.1), the system can then be represented by a two-state Markov chain with just two states: $\omega^1$, where every player adopts strategy $s_1$, and $\omega^2$ where nobody does.

The following question is now asked: What is the long-run behaviour of the system when the probability of experimentation is small? First, it has to be verified that such a notion as *the* long-run behaviour of the system is well defined in the present context (in particular, unique, independently of initial conditions). But this is an immediate consequence of the fact that, due to mutation, there is positive probability no lower than $\epsilon^n$ (associated with the event that every individual simultaneously mutates) for a transition across the two states (in either direction) to materialize every period. Thus, from the well-known Ergodic Theorem (see, for example, Karlin and Taylor (1975) for a standard review of the Theory of Stochastic Processes), the Markov chain has a unique invariant distribution $\mu(\cdot)$ which summarizes the long-run behaviour of the system in the following strong sense: with *ex ante* probability one, the frequency with which each of the two states, $\omega^1$ and $\omega^2$, is visited along any sample path of the process converges to $\mu(\omega^1)$ and $\mu(\omega^2)$ respectively.

Of course, the particular form of this unique invariant distribution must depend on the mutation probability rate $\epsilon$ under consideration. Thus, write $\mu_\epsilon$ to express such dependence. Then, more precisely, the previous question can be reformulated as follows. What is the limit of $\mu_\epsilon$ when $\epsilon \to 0$?

To compute the limit of $\mu_\epsilon$,[4] it is enough to rely on the following bounds on the cross-transition probabilities of the process.

---

[4] Again, first of all, one needs to make sure that this limit is well defined (i.e. unique). But this may be confirmed by simply observing that each entry in the transition matrix of the Markov chain is a polynomial in $\epsilon$.

On the one hand, there exists some $\bar{\epsilon} > 0$, $\beta > 0$ such that, if $0 < \epsilon \leq \bar{\epsilon}$,

$$q_{21} \equiv \Pr\{\omega_{t+1} = \omega^1 \mid \omega_t = \omega^2\} \geq \beta \epsilon^2. \tag{5.2}$$

To verify this, simply note that, if at some $t$ every player is planning to play strategy $s_2$ (prior to mutation), it is enough that two players mutate towards $s_1$ *and* become paired by the matching mechanism for the transition towards state $\omega^1$ to take place. (Then, the average pay-off of $s_1$ will be higher than that of $s_2$, which by (5.1) leads to such transition.) For small $\epsilon$, this event is clearly of order $\epsilon^2$.

Assume now that the population is large enough. (Specifically, it is enough that $n \geq 8$.) Then, there exists some $\tilde{\epsilon} > 0$, $\gamma > 0$ such that, if $0 < \epsilon \leq \tilde{\epsilon}$, the converse transition probability

$$q_{12} \equiv \Pr\{\omega_{t+1} = \omega^2 \mid \omega_t = \omega^1\} \leq \gamma \epsilon^3. \tag{5.3}$$

The reason here is that, if at some $t$ every player is planning to play strategy $s_1$ (prior to mutation), then *only two* players mutating is *not* enough (under *any* realization of the matching mechanism) for strategy $s_2$ to enjoy a higher average pay-off than $s_1$. Therefore, at least three mutations are required for a transition to $\omega^2$ to take place which, for small $\epsilon$, is an event of order no higher than $\epsilon^3$.

Since the invariant distribution of the two-state Markov chain is given by

$$\mu_\epsilon(\omega^2) = \frac{q_{12}}{q_{12} + q_{21}}$$
$$\mu_\epsilon(\omega^1) = 1 - \mu_\epsilon(\omega^2)$$

a combination of (5.2) and (5.3) yields:

$$\lim_{\epsilon \to 0} \mu_\epsilon(\omega^2) = 0$$
$$\lim_{\epsilon \to 0} \mu_\epsilon(\omega^1) = 1.$$

That is, in the long run, for small mutation probability, the process spends most of the time at the state where all players play strategy $s_1$, the efficient strategy.[5]

Consider now Scenario 2. In this case, the average pay-off earned by each strategy can be identified with its expected pay-off, due to the Law of Large Numbers and the postulated independence across matching rounds taking place within each period (see Section 5.3 for a precise formulation of this idea). The important point to notice here is that, unlike Scenario 1, these expected/average magnitudes only depend on the current number of individuals $(z_t, n - z_t)$ who play each strategy. Adapting previous notation, they are defined as follows (of course, provided $0 < z_t < n$):

$$\hat{\pi}_1(z_t) = \frac{\sqrt{3}(z_t - 1)}{n - 1};$$
$$\hat{\pi}_2(z_t) = 1.$$

[5] Strictly speaking, it is only shown that players will spend most of the time "planning" to play $s_1$ (i.e. before mutation). However, since the mutation probability is taken to be small, they will also end up actually playing $s_1$ most of the time.

On the basis of these pay-offs and the maintained simplifying assumption that players adjust their planned strategies immediately, the analogue of (5.1) for this case may be written as follows:

$$\hat{B}(z_t) = \begin{cases} n & \text{if} \quad \hat{\pi}_1(z_t) > \hat{\pi}_2(z_t) \\ 0 & \text{if} \quad \hat{\pi}_1(z_t) < \hat{\pi}_2(z_t). \end{cases} \tag{5.4}$$

Focusing again on the states of the system right before mutation takes place in each period, denote by $\hat{\mu}_\epsilon(\cdot)$ the corresponding invariant distribution when the mutation rate is $\epsilon > 0$. To evaluate it when $\epsilon \to 0$, first note that

$$\hat{\pi}_1\left(\frac{n}{2}\right) < \hat{\pi}_2\left(\frac{n}{2}\right). \tag{5.5}$$

Therefore, there exists some $\hat{\epsilon} > 0$, $\hat{\beta} > 0$, $\hat{\gamma} > 0$, such that, if $0 < \epsilon < \hat{\epsilon}$, the following two inequalities apply:

$$q_{21} \equiv \Pr\left\{\omega_{t+1} = \omega^1 \mid \omega_t = \omega^2\right\} \leq \hat{\beta}\,\epsilon^{\frac{n}{2}+1} \tag{5.6}$$

and

$$q_{12} \equiv \Pr\left\{\omega_{t+1} = \omega^2 \mid \omega_t = \omega^1\right\} \geq \hat{\gamma}\,\epsilon^{\frac{n}{2}}. \tag{5.7}$$

The first inequality is a consequence of the fact that, from (5.4) and (5.5), at least $\frac{n}{2} + 1$ mutations are needed to trigger a transition from $\omega^2$ to $\omega^1$. The second inequality also follows from (5.4)–(5.5) which jointly imply that, reciprocally, $\frac{n}{2}$ mutations are sufficient for a transition from $\omega^1$ to $\omega^2$ to occur.

A combination of (5.6) and (5.7) yields

$$\lim_{\epsilon \to 0} \hat{\mu}_\epsilon(\omega^1) = 0$$
$$\lim_{\epsilon \to 0} \hat{\mu}_\epsilon(\omega^2) = 1.$$

That is, in the long run, for small mutation probability, the process selects the opposite strategy as before.

The previous example serves two basic purposes.

First, it illustrates the essential "mutation-counting logic" which, in a less stylized form, will underlie much of our future analysis. In general, the task of finding the long-run (stochastically stable) states will be seen to be equivalent to identifying those states whose "stochastic potential" (heuristically, the difference between the number of mutations needed to reach and exit them) is maximal.

The previous example also makes clear the following second point. Certain details of the environment (for example, the relative importance to be attributed to the different sources of noise) will generally have an interesting and crucial role to play in the model's long-run predictions.

## 5.3   Theoretical Framework

Consider now the following theoretical framework. As before, there is a finite population of $n$ individuals which are randomly matched every period $t = 1, 2, ...$, to play a certain bilateral symmetric game. Within each period, there is some fixed number $v \in \mathcal{N}$ of independent matching rounds, with each player adopting the same action in every one of them.

For the moment, attention is restricted to games with just two strategies, represented by the following general pay-off table:

|       | $s_1$ | $s_2$ |
|-------|-------|-------|
| $s_1$ | $a, a$ | $d, c$ |
| $s_2$ | $c, d$ | $b, b$ |

Table 15

Generalizations of this two-strategy framework will be discussed below in Subsection 5.4.4.

Within such a $(2 \times 2)$-scenario, the most interesting case corresponds to a game which is of the co-ordination type, i.e. both $(s_1, s_1)$ and $(s_2, s_2)$ are strict equilibria. This amounts to the following restrictions on pay-offs:

$$a > c, \, b > d, \tag{5.8}$$

which will be maintained throughout most of this chapter. The other two alternative cases which (generically) exhaust all other possibilities are given by:

$$(a - c)(d - b) > 0, \tag{5.9}$$

and

$$a < c, \, b < d. \tag{5.10}$$

The first case reflects a game where one of the strategies strictly dominates the other. The second case involves a game whose unique symmetric equilibrium involves mixed strategies (specifically, each player chooses strategy $s_1$ with probability $\gamma = \frac{d-b}{c-a+d-b}$). As mentioned, (5.8) raises equilibrium-selection issues which are much more interesting than those induced by (5.9) or (5.10), where the issue of equilibrium selection is trivially absent. Therefore, these two latter cases will be only informally discussed below.

The state space of the process is identified with the set

$$\Omega = \{\omega = (z, n - z) : z \in \mathcal{N}, \, 0 \le z \le n\},$$

where, at each $t$, $\omega_t = (z_t, n - z_t) \in \Omega$ is interpreted as the number of individuals $z_t$ and $n - z_t$ respectively *playing* strategies $s_1$ and $s_2$ at $t$.[6] Along the lines proposed for the above example, the state of the system will change in response to two different mechanisms: selection and mutation.

First, the selection component of the dynamics is described. Formulated in a very flexible manner, it is simply required to satisfy the requirement that the frequency of any strategy which uniquely enjoys the highest pay-off should never decrease, displaying also some positive probability of strictly increasing. This reflects a very weak notion of monotonicity, akin to those considered in previous chapters.

To express it formally, let $\tilde{p}_t^v$ denote the random variable (a function of $\omega_t$) expressing the number of cross-pairings prevailing at $t$ along $v$ independent matching rounds. Associated with any realization $p_t^v$ of this random variable, the average pay-offs obtained by each strategy are given by:

$$\pi_1(z_t, p_t^v) = \frac{[a(v z_t - p_t^v) + d p_t^v]}{v z_t},$$

$$\pi_2(z_t, p_t^v) = \frac{[c p_t^v + b(v(n - z_t) - p_t^v)]}{v(n - z_t)}, \tag{5.11}$$

provided, of course, both strategies are played in positive frequency, i.e. $z_t > 0$ and $n - z_t > 0$. Note of course that, *ex ante*, these average pay-offs are typically random since the number of cross-pairings defines a random variable which generally has a non-degenerate support (specifically, this happens when $1 < z_t < n - 1$).

Given $z_t$ and $p_t^v$ (respectively, the number of $s_1$-adopters and total number of cross-matchings prevailing at $t$) let $\tilde{B}(z_t, p_t^v) \in \Omega$ denote the random variable which determines the new number of $s_1$-adopters after strategy adjustment. This random variable determines the state of the process at $t+1$ *prior* to the operation of mutation. The previous verbal requirement on the selection dynamics may be formalized as follows:

$$\tilde{B}(z_t, p_t^v) \begin{cases} \geq z_t & \text{if } \pi_1(z_t, p_t^v) > \pi_2(z_t, p_t^v) \\ \leq z_t & \text{if } \pi_1(z_t, p_t^v) < \pi_2(z_t, p_t^v), \end{cases} \tag{5.12}$$

where all these inequalities are required to be strict with some positive probability. In case $\pi_1(z_t, p_t^v) = \pi_2(z_t, p_t^v)$, the dynamics may be specified in any arbitrary manner, inessential to the analysis.

Note that this formulation admits of a large variety of concrete specifications for the adjustment process; in particular, it allows for a substantial degree of inertia in strategy adjustment. The only essential requirement is that when adjustment takes place (which occurs with some probability unless both strategies yield identical average pay-offs), it is never performed against (and sometimes towards) the strategy with the highest average pay-off. As before, the

---

[6] Thus, unlike in the example discussed above, the state at $t$ reflects both the strategy adjustment *and* mutation occurring between $t - 1$ and $t$.

two monomorphic configurations ($z_t = 0$ or $z_t = n$) will be assumed stationary from the viewpoint of the selection dynamics, i.e. $\tilde{B}(0,0) = 0$ and $\tilde{B}(n,0) = n$.

Selection is the core component of the evolutionary dynamics. With a very high probability, transitions take place exclusively in accordance with it. Occasionally, however, it is perturbed by mutation, a phenomenon which in social environments may be interpreted as formalizing some notion of, say, individual experimentation or population renewal. In this respect, we adopt the formulation described in the previous section and assume that, prior to play being conducted at $t+1$ and after selection has modified the former configuration prevailing at $t$, every individual is subject to some independent (and, for simplicity, common) mutation probability $\epsilon > 0$. In the event of mutation, each agent is assumed to adopt *any* of the available strategies with some positive (say, equal) probability.

When the phenomenon of mutation is added to the selection dynamics, every transition has positive probability. Therefore, the stochastic process becomes an aperiodic and irreducible Markov chain on the state space $\Omega$. This, in turn, has the important analytical consequence explained above: the process has a unique invariant distribution $\mu \in \Delta(\Omega)$ which fully summarizes its long-run behaviour, almost surely (a.s.), independently of initial conditions. To express the dependence of this invariant distribution $\mu$ on the key parameters of the model, $\epsilon$ and $v$, the notation $\mu_{\epsilon,v}(\cdot)$ will be used.

As suggested by the simple example discussed in Section 5.2, one of the interesting features to be expected from the evolutionary framework proposed is a rich interaction between its different sources of noise: mutational and matching-induced noise. Indeed, as the next section analyses in detail, different relative magnitudes for each of them (as they become small) lead to sharply different conclusions on the long-run selection achieved by the model.

## 5.4 Analysis

In order to illustrate in a clear-cut way the different issues involved, the analysis will be structured in two different polar scenarios. In the first one, the mutational noise is of a smaller order of magnitude than the matching-induced noise. This asymmetry arises because, fixing the number of matching rounds per period (arbitrarily large but finite), the analysis focuses on the long-run behaviour of the system when the mutation probability becomes infinitesimal (i.e. converges to zero).

The second scenario considers the opposite case. First, it removes all matching uncertainty by having the number of matching rounds per period become unboundedly large for any *given* mutation rate. Once matching uncertainty is so eliminated (i.e. expected and average magnitudes coincide a.s. due to the Law of Large Numbers), the analysis focuses on the long-run performance of the model for an arbitrarily small mutation probability.

Formally, these two polar scenarios are associated with two different orders in which the limit operations on the parameters $\epsilon$ and (possibly) $v$ are conducted. In the former case, the limit on $\epsilon$ is taken first; in the second case, on the contrary, it is the limit as $v \to \infty$ which is computed initially. Focusing on each of these two alternative scenarios, one obtains a stylized (i.e. extreme) *approximation* for the state of affairs prevailing when both the mutation probability and the matching noise become small, but at different relative rates. A conceptual motivation for this theoretical exercise is provided below (see Subsection 5.4.3), once the corresponding formal analysis has been completed.

### 5.4.1    Large matching noise

Define

$$\mu_v^*(\cdot) \equiv \lim_{\epsilon \to 0} \mu_{\epsilon,v}(\cdot), \tag{5.13}$$

i.e. the limit invariant distribution obtained, for any given $v$, when the mutation probability becomes arbitrarily small. The next result shows that $\mu_v^*$ is a well-defined element of $\Delta(\Omega)$ which is independent of the particular value of $v$ considered. Moreover, it uniquely selects the efficient outcome, provided the population is large enough.

**Theorem 19 (Robson and Vega-Redondo (1996))** *Assume (5.8), $a > b$, and let $\omega^1 \equiv (n, 0)$ be the state where all players adopt $s_1$ (the efficient strategy). For all $v \in \mathcal{N}$, $\mu_v^*(\cdot) \in \Delta(\Omega)$. Moreover, there exists some $\hat{n} \in \mathcal{N}$ such that if $n \geq \hat{n}$, then $\mu_v^*(\omega^1) = 1$.*

**Proof.** Given $v$, the number of matching rounds per period, let $T_\epsilon$ stand for the transition matrix of the evolutionary process described, where $\epsilon$ is the corresponding mutation rate and $T_\epsilon(\omega, \omega')$ indicates the probability that the process transits from state $\omega$ to state $\omega'$. As explained, for $\epsilon > 0$, the existence of an invariant distribution $\mu_{\epsilon,v}(\cdot)$ is guaranteed by the ergodicity of the induced stochastic process.

Now, we argue that the limit invariant distribution defined by (5.13) is also well defined. This follows directly from the graph characterization described in the Appendix. By (7.1) and Proposition 17, each of the components of $\mu_{\epsilon,v}(\cdot)$, for any $\epsilon > 0$ and $v \in \mathcal{N}$, is given by a ratio of polynomials in $\epsilon$. Therefore, its limit is well defined when $\epsilon \to 0$.

As a further consequence of Proposition 17, one may also conclude that the support of $\mu_v^*(\cdot)$ – i.e. the set of *stochastically stable states* – consists of those $\omega$ whose corresponding $q_\omega$ display, as $\epsilon \to 0$, the smallest rate of convergence to zero, $\mathcal{O}(q_\omega)$. Since every $q_\omega$ is a polynomial in $\epsilon$, $\mathcal{O}(q_\omega)$ coincides with its lowest exponent of $\epsilon$. This, in turn, from mere inspection of (7.1), is seen to be equal to the *minimum* total number of mutations required in some $\omega$-tree $Y \in \mathcal{Y}_\omega$.

To make this discussion precise, it is useful to resort to a certain notion of cost associated with the different possible transitions. Denote by $T_0$ the transition matrix associated with the mutation-free dynamics (i.e. the evolutionary dynamics resulting when $\epsilon = 0$). Furthermore, define a distance function on $\Omega$, $d : \Omega \times \Omega \rightarrow \mathcal{N}$, which for every pair of states, $\omega = (z, n - z)$ and $\omega' = (z', n - z')$, indicates the minimum number of agents who must be playing a different action at $\omega$ and $\omega'$, i.e.

$$d(\omega, \omega') = |z - z'| = |(n - z) - (n - z')|.$$

Then, a cost function $c : \Omega \times \Omega \rightarrow \mathcal{N}$ is defined as follows:

$$c(\omega, \omega') = \min_{\omega'' \in \Omega} \{ d(\omega'', \omega') : T_0(\omega, \omega'') > 0 \}. \tag{5.14}$$

This cost function indicates the minimum number of mutations required to complement the mutation-free dynamics in order to carry out any contemplated transition. This cost function can be extended to any path $y$ or tree $Y$ in the natural fashion: the cost of a path or a tree is identified with the aggregate cost of its constituent links. Extending the former notation, such a cost is denoted by $c(y)$ or $c(Y)$.

From the considerations explained above, it follows that, as $\epsilon \rightarrow 0$, the order $\mathcal{O}(q_\omega)$ for any given state $\omega$ is equal to the minimum cost $c(Y)$ achievable for some $Y \in \mathcal{Y}_\omega$. Thus, combining this with Proposition 17, one concludes that the task of identifying the stochastically stable states simply reduces to determining which states $\omega \in \Omega$ display a corresponding $\omega$-tree $Y$ of minimum cost, i.e. a tree $Y \in \mathcal{Y}_\omega$ whose cost $c(Y)$ is no larger than any other $Y' \in \mathcal{Y}_{\omega'}$ for all $\omega' \in \Omega$.

This search is substantially facilitated by the following two Lemmas. To state them precisely, one needs the following standard definition, as particularized to the present framework.

**Definition 19** *A certain $D \subseteq \Omega$ is a limit (or absorbing) set of the mutation-free dynamics induced by $T_0$ if:*
*(a) $\forall \omega \in D, T_0(\omega, \omega') > 0 \Rightarrow \omega' \in D$;*
*(b) $\forall \omega, \omega' \in D, \exists m \in \mathcal{N}$ such that $T_0^{(m)}(\omega, \omega') > 0$.*[7]

**Lemma 3** *Let $\omega \in supp(\mu_v^*)$.*[8] *Then, $\omega$ belongs to some limit set of the mutation-free dynamics.*

*Proof:* For any given $\epsilon > 0$, the invariant distribution $\mu_{\epsilon,v}$ satisfies:

$$\forall \omega \in \Omega, \quad \sum_{\omega' \in \Omega} \mu_{\epsilon,v}(\omega') \, T_\epsilon(\omega', \omega) = \mu_{\epsilon,v}(\omega).$$

---

[7] Here, $T_0^{(m)}$ stands for the $m$th iteration of the transition matrix $T_0$.
[8] As customary, the notation $supp(\cdot)$ stands for the support of the probability distribution in question.

By taking limits on the above expression for $\epsilon \to 0$, and relying on the continuity of $T_\epsilon$ in $\epsilon$, one obtains

$$\forall \omega \in \Omega, \quad \sum_{\omega' \in \Omega} \mu_v^*(\omega') \, T_0(\omega', \omega) = \mu_v^*(\omega).$$

This expression obviously implies that if $\mu_v^*(\omega) > 0$, it must also belong to a limit set of the mutation-free dynamics induced by $T_0$, thus completing the proof of the Lemma.

**Lemma 4** *There exists some $\tilde{n} \in \mathcal{N}$ such that if $n \geq \tilde{n}$, the only limit sets of the mutation-free dynamics are the singletons $\{\omega^1\}$ and $\{\omega^2\}$, where $\omega^i$ stands for the state where every agent adopts $s_i$.*

*Proof:* Define the (mutation-free) basin of attraction $U^i$ of the limit set $\{\omega^i\}$, $i = 1, 2$, as the subset of the state space from which there is a positive probability to make a transition to this limit set in some finite number of periods *without* relying on mutation. Formally,

$$U^i \equiv \{\omega \in \Omega : \exists m \in \mathcal{N} : T_0^{(m)}(\omega, \omega^i) > 0\}.$$

Now choose $k$ as the smallest integer such that:

$$\frac{a(k-1) + d}{k} > \max\{b, c\} \equiv f. \tag{5.15}$$

Note that $k$ depends only on the pay-offs of the game, and not on $n$, the number of players. The proof of the Lemma is based on the following two Claims.

*Claim 1:* If $\omega = (z, n - z)$ satisfies $z \geq k$, then $\omega \in U^1$.

To prove this claim, note that if there are at least $k$ individuals playing strategy $s_1$, then, provided all of them meet among themselves in each round of matching (with the possible exception of one of them, if $z$ is an odd number), strategy $s_1$ will earn an average pay-off higher than $s_2$. Since this matching outcome displays some positive probability, the same applies to having the ensuing state involve at least one more player adopting strategy $s_1$. Proceeding in this fashion for at most $n - z$ periods, the stated conclusion follows.

The second claim pertains to the basin of attraction of the alternative limit state $\omega^2$.

*Claim 2:* There exists some $\tilde{n} \in \mathcal{N}$ such that if $n \geq \tilde{n}$ and $\omega = (z, n - z)$ satisfies $z < k$, then $\omega \in U^2$.

The proof of this second claim derives from the fact that if $n$ is large enough and fewer than $k$ players are adopting strategy $s_1$, the average pay-off of strategy $s_1$ is bound to be lower than that of $s_2$, provided all individuals playing the former strategy are matched with individuals playing the latter in every matching round. For in this case, the average pay-off of $s_2$ will be arbitrarily close to $b$ if $n$ is large enough. And this average pay-off will be larger than $d$, which is the pay-off obtained in *every* one of their encounters by players adopting strategy $s_1$.

From Claims 1 and 2, there is positive probability that, from any state $\omega \in \Omega$, the mutation-free dynamics will become absorbed by either $\omega^1$ or $\omega^2$ in some finite number of steps. This implies, as desired, that $\{\omega^1\}$ and $\{\omega^2\}$ are the only limit sets of the mutation-free dynamics.

The proof of the Theorem is then completed through the following two Lemmas.

**Lemma 5** *Let $n \geq \tilde{n}$, as in Lemma 4. Then there exists an $\omega^1$-tree $Y$ such that $c(Y) = k$, where $k$ satisfies (5.15).*

*Proof:*   Let $\tilde{\omega} = (k, n - k)$ and consider the path $\tilde{y} = \{(\omega^2, \tilde{\omega}), (\tilde{\omega}, \omega^{(1)}), (\omega^{(1)}, \omega^{(2)}), ..., (\omega^{(r)}, \omega^1)\}$ where $T_0(\tilde{\omega}, \omega^{(1)}) > 0$, $T_0(\omega^{(r)}, \omega^1) > 0$, and $T_0(\omega^{(q)}, \omega^{(q+1)}) > 0$ for all $q = 1, 2, ..., r - 1$. From Claim 1 above, such a path can be constructed, with $c(\tilde{y}) = k$. Moreover, since all states in $\Omega \backslash \{\omega^1, \omega^2\}$ belong to either $U^1$ or $U^2$ (cf. again Claims 1 and 2 above) it is clear that the path $\tilde{y}$ can be completed to form a $\omega^1$-tree without increasing the cost of $k$ contributed by this path. This completes the proof of the Lemma.

**Lemma 6** *There exists some $\hat{n}$ ($\geq \tilde{n}$ in Lemma 4) such that if $n \geq \hat{n}$, then $\min \{c(Y) : Y \in \mathcal{Y}_{\omega^2}\} > k$, where $k$ is chosen as in Lemma 5.*

*Proof:*   Choose $\hat{n} \in \mathcal{N}$ to satisfy

$$\hat{n} > \max \left\{ \frac{(2a - d - f)k}{a - f}, \tilde{n} \right\},$$

where $f$ is defined in (5.15). If $n \geq \hat{n}$, then $\frac{n}{2} > k$, and

$$\frac{a(n - 2k) + dk}{n - k} > f. \tag{5.16}$$

Consider now any $\omega = (z, n - z)$ where $z \geq n - k$. By (5.16), the average pay-off obtained by strategy $s_1$ must always be larger than that of $s_2$ for *any* outcome of the matching process. In particular, therefore, one has that $\omega \notin U^2$.[9]

---

[9] Of course, if $\omega \notin U^2$ it must be the case that $\omega \in U^1$. However, notice that, in general, the latter statement does *not* imply the former since, in the evolutionary dynamics considered, basins of attraction are not disjoint sets.

Let $Y$ be any given $\omega^2$-tree. As required by Definition 21, this tree must include a path $y$ from $\omega^1$ to $\omega^2$. From the considerations above, $c(y) > k$. Therefore, $c(Y) > k$, which completes the proof of the Lemma. ∎

Since the conclusion of Theorem 19 is independent of $v$, the number of matching rounds per period, the next obvious Corollary follows.

**Corollary 5** *Under the conditions of Theorem 19, there exists some $\hat{n} \in \mathcal{N}$ such that if $n \geq \hat{n}$, then*

$$\lim_{v \to \infty} \lim_{\epsilon \to 0} \mu_{\epsilon,v}(\omega^1) = 1. \qquad (5.17)$$

### 5.4.2    Small matching noise

Now we consider the polar scenario where, in taking limits on $v$ and $\epsilon$, matching noise is made to vanish at a faster pace than mutational noise. This amounts to interchanging the order of the limit operation in (5.17), giving rise to the following result.

**Theorem 20 (Kandori** *et al.* **(1993))** *Assume (5.8), $c+b > a+d$, and let $\omega^2 \equiv (0, n)$ be the state where all players adopt $s_2$ (the risk-dominant strategy).*[10] *There exists some $\tilde{n} \in \mathcal{N}$ such that if $n \geq \tilde{n}$, then*

$$\lim_{\epsilon \to 0} \lim_{v \to \infty} \mu_{\epsilon,v}(\omega^2) = 1. \qquad (5.18)$$

**Proof.** Given any $\epsilon > 0$, denote by $\hat{T}_\epsilon$ the transition matrix of the (limit) evolutionary process resulting when all matching uncertainty is removed by making $v \to \infty$. In this case, (5.12) can be rewritten as:[11]

$$\hat{B}(z_t) \begin{cases} \geq z_t & \text{if } \hat{\pi}_1(z_t) > \hat{\pi}_2(z_t) \\ \leq z_t & \text{if } \hat{\pi}_1(z_t) < \hat{\pi}_2(z_t) \, , \end{cases} \qquad (5.19)$$

where $\hat{B}(z_t)$ stands for the random variable expressing the number of new $s_1$-adopters after strategy adjustment at $t + 1$ (prior to mutation) and each $\hat{\pi}_i(z_t)$ stands for the expected pay-off obtained at $t$ by each strategy $s_i$, $i = 1, 2$. By

---

[10] The well-known concept of *risk-dominance* has been proposed by Harsanyi and Selten (1988). Of course, risk-dominance and Pareto efficiency may well conflict as illustrated e.g. by the game discussed in Section 5.2.

[11] Notice that, as before, these inequalities must continue to hold strictly with some positive probability, provided that (5.12) does not depend directly on $v$ (as was implicitly postulated). In this case, (5.19) must inherit such a requirement from (5.12).

the Law of Large Numbers, these expected pay-offs equal average pay-offs a.s. and are given by:

$$\hat{\pi}_1(z_t) \equiv E(\pi_1(z_t, \tilde{p}_t^v)) = \frac{a(z_t - 1) + d(n - z_t)}{n - 1},$$

$$\hat{\pi}_2(z_t) \equiv E(\pi_2(z_t, \tilde{p}_t^v)) = \frac{cz_t + b(n - z_t - 1)}{n - 1}.$$

Denote $\hat{\mu}(\cdot) \equiv \lim_{\epsilon \to 0} \lim_{v \to \infty} \mu_{\epsilon,v}(\cdot)$. From considerations identical to those explained in the proof of Theorem 19, this limit is a well-defined probability distribution on $\Omega$. In characterizing its support, one may rely again on the direct counterpart of Lemma 3 which, for completeness, is now stated without proof.

**Lemma 7** *Let $\omega \in supp(\hat{\mu})$. Then $\omega$ belongs to some limit set of the mutation-free dynamics induced by $\hat{T}_0$.*

As before, the next step is to characterize the limit sets of the mutation-free dynamics. This is the content of the following obvious Lemma, again stated without proof.

**Lemma 8** *Let $\rho$ be the unique real number that satisfies $a(\rho - 1) + d(n - \rho) = c\rho + b(n - \rho - 1)$.[12] The only limit sets of the mutation-free dynamics are $\{\omega^1\}$, $\{\omega^2\}$ and, possibly (if $\rho \in \mathcal{N}$), the singleton $\{\tilde{\omega}\}$ where $\tilde{\omega} = (\rho, n - \rho)$.*

Define now a cost function $\hat{c} : \Omega \times \Omega \to \mathcal{N}$, which is the direct counterpart of that defined in (5.14), i.e.

$$\hat{c}(\omega, \omega') = \min_{\omega'' \in \Omega} \{d(\omega'', \omega') : \hat{T}_0(\omega, \omega'') > 0\}.$$

From the considerations already explained in the proof of Theorem 19, the present result follows from the following Lemmata.

**Lemma 9** *There exists some $\tilde{n} \in \mathcal{N}$ such that if $n \geq \tilde{n}$, an $\omega^2$-tree $Y$ exists with $\hat{c}(Y) \leq \frac{n}{2}$.*

*Proof:* Let $\rho$ be defined as in the statement of Lemma 8. From the fact that $a + d < c + b$, there must exist some $\tilde{n} \in \mathcal{N}$ such that if $n \geq \tilde{n}$, the corresponding value of $\rho > \frac{n}{2}$. Which, in particular, implies that $\hat{\pi}_2(\frac{n}{2}) > \hat{\pi}_1(\frac{n}{2})$, i.e. when exactly half of the population plays each strategy, the average pay-off of strategy $s_2$ is higher than that of $s_1$.

For concreteness, assume that $\rho$ is an integer so that the state $\tilde{\omega} = (\rho, n - \rho)$ is well defined. (If it were not defined, then the ensuing argument would merely

---

[12] Thus, if $\rho$ happens to be an integer, it is the precise number of $s_1$-adopters for which $\hat{\pi}_1(\rho) = \hat{\pi}_2(\rho)$. In general, however, it will not be an integer and will partition the state space into the basins of attraction of each monomorphic state, $\hat{U}^1 = \{\omega = (z, n - z) : z > \rho\}$ and $\hat{U}^2 = \{\omega = (z, n - z) : z < \rho\}$.

be simplified.) By the definition of $\rho$, the state $\hat{\omega} = (\rho - 1, n - \rho + 1)$ belongs to the basin of attraction of $\omega^2$. Therefore, there is a path $y$ joining $\hat{\omega}$ and $\omega^2$ at zero cost. Since $\hat{c}(\tilde{\omega}, \hat{\omega}) = 1$ and

$$\hat{c}(\omega^1, \tilde{\omega}) = n - \rho \le \frac{n}{2} - 1,$$

the path $y' = \{(\omega^1, \tilde{\omega}), (\tilde{\omega}, \hat{\omega})\} \cup y$ joins $\omega^1$ to $\omega^2$ at a total cost

$$\hat{c}(y') \le \frac{n}{2}.$$

Moreover, $y'$ includes in it all three limit states, $\omega^1$, $\omega^2$, and $\tilde{\omega}$. Therefore, it can be completed to construct a $\omega^2$-tree at no additional cost. This completes the proof of the Lemma. ∎

**Lemma 10** *There exists some $\tilde{n} \in \mathcal{N}$ such that if $n \ge \tilde{n}$, then $\min \{\hat{c}(Y) : Y \in (\mathcal{Y}_{\omega^1} \cup \mathcal{Y}_{\tilde{\omega}})\} > \frac{n}{2}$.*

*Proof:*   Assume again, for the sake of concreteness, that $\tilde{\omega}$ is well defined. To show that every tree belonging to either $\mathcal{Y}_{\omega^1}$ or $\mathcal{Y}_{\tilde{\omega}}$ must have a cost larger than $\frac{n}{2}$, it is enough to show that all of them must have some path with this property. Specifically, note that every such tree must include a path $y$ which joins $\omega^2$ to some state $\omega = (z, n - z)$ with $z \ge \rho$, where $\rho$ is as defined above. It is now argued that this path must satisfy $\hat{c}(y) \ge \frac{n}{2} + 1$. First, notice that, for sufficiently large $n$, $\rho > \frac{n}{2}$. Therefore, any state $\omega'$ along path $y$ for which the subpath joining $\omega^2$ and $\omega'$ has a cost no larger than $\frac{n}{2}$ must still belong to the basin of attraction of $\omega^2$. Since in the present context (with no matching noise), all basins of attraction are disjoint, this implies that further mutations are needed to reach state $\omega^1$ with positive probability. This completes the proof. ∎

### 5.4.3    On the role of noise in evolutionary models

The analysis just presented illustrates how the addition of genuine stochastic noise to an evolutionary framework may fruitfully interact with the forces of selection to provide a strong and clear-cut selection device in games. Even in those cases where the game has several *strict* Nash equilibria (and, therefore, all of them are immune to any of the usual criteria of Nash refinement) this evolutionary approach is able to pinpoint a particular equilibrium. Specifically, it is the one associated with the unique (stochastically stable) state remaining within the support of the limit invariant distribution as the noise of the system becomes vanishingly small. In general, the equilibrium selected reflects both the pay-off structure of the game *and* the relative orders in which the different sources of noise are assumed to vanish.

In contrast with the developments discussed in previous chapters, the key new feature of the present approach is the role played by noise (both matching noise and mutation) in liberating the process from absorption at one of the different limit sets of the selection dynamics. The introduction of this new evolutionary factor raises a number of interesting (and somewhat controversial) issues which are worth elaborating upon in some detail in what follows.

Each of the two sources of noise contemplated by the evolutionary process plays a quite different role in the underlying logic of the model.

Mutational noise, on the one hand, is best viewed as a source of perturbation of the "core" (i.e. selection) component of the process. In a sense, it entails a test of robustness for each of the several limit sets which will typically result from the exclusive operation of the selection (mutation-free) dynamics.[13] Allowing for mutation, one may discriminate among all these sets, based on the fact that not all of them are equally robust under mutation. Only those which are most resistant to mutation may qualify as long-run outcomes. They will include the most frequently observed states when, in the long run, mutation is given enough time to upset the temporary workings of the selection dynamics.[14]

Matching noise, on the other hand, fulfils a very different (but complementary) role in the evolutionary process. By itself, it can never upset a configuration where, for example, the population is fully co-ordinated at a certain equilibrium. However, when remaining at a significant level (i.e. the case addressed by Theorem 19), it enhances the power and flexibility by which mutations (even when appearing in small numbers) may impinge on the long-run evolution of the process.[15] But, in general, matching-induced noise will not just be a mere unbiased catalyst. As exemplified by our former analysis, it may well display some inherent biases which *qualitatively* affect the long-run behaviour of the process.

Besides their "technical" usefulness as a selection device, one may still wonder what is the *conceptual* basis for including these different types of noise in the evolutionary framework. Concerning mutation, it has already been amply explained that it represents an essential building block of evolutionary processes and, as such, is best modelled in a fully fledged manner. Mutation provides the indispensable source of variation needed for selection forces to operate fruitfully. In this sense, the introduction of mutational noise as an *explicit* component of the theoretical framework should be seen as addressing one of the main conceptual voids left untackled by previous developments.

The motivation for allowing for a significant extent of noise to persist within

[13] Recall e.g. Lemmata 4 and 8 above. One trivial exception to this multiplicity of limit sets occurs when one of the strategies strictly dominates all others. In this case, and provided the number of players is large enough, the only non-degenerate limit set has all players choosing the dominant strategy. As explained in Subsection 5.4.4, an analogous situation happens when the game has a unique (mixed-strategy) equilibrium.

[14] As formulated, the perturbation on the selection dynamics induced by mutation is unbiased in two implicit (but crucial) respects. First, it may lead to *any* ensuing strategy with non-vanishing probability when it occurs. Second, its (full) probability of occurrence is state-independent. The first condition is obviously required if the above analysis is to apply (otherwise, the process may be non-ergodic). The second condition is also essential, as shown by Bergin and Lipman (1995).

[15] See Section 5.6 for a precise formalization of the role of matching noise in increasing the rate of convergence of the process.

the matching framework is somewhat more controversial. One could probably argue that evolutionary models are only suitably applied to large populations and that, therefore, matching noise (or, for that matter, any other source of noise associated with relatively small populations) is best ignored. However, much in line with analogous ideas put forward by some biologists (most notably, Sewal Wright – see Wright (1945)), one could also argue that occasional (perhaps quite unlikely) "small-population effects" (i.e. effects which cannot always be discarded by resorting to the Law of Large Numbers) are crucial to understanding the evolution of some beneficial traits in large populations.[16]

In the evolutionary framework described, the relative magnitudes associated with the two alternative sources of noise are formally captured by the particular order in which the limit operations are sequentially taken (first on $\epsilon$ then on $v$, or vice versa).[17] As the previous discussion suggests, the "right" order in which the different limit operations are to be taken must be tailored to the particular details of the context under consideration. In general, it will have to be one of the choices made at the "modelling stage" in the analysis.

### 5.4.4　Extensions

Theorems 19 and 20 have focused on symmetric games with just two strategies and two different co-ordination equilibria. As mentioned, this is the case which in the context of $(2 \times 2)$-games raises the most interesting questions, and where evolutionary analysis makes a more substantive contribution to traditional game-theoretic analysis. For the other two kinds of symmetric $(2 \times 2)$-games which generically cover all other possibilities (those given by inequalities (5.9) and (5.10)), the game has a unique equilibrium. In the first case, this (pure strategy) equilibrium is trivial since one of the strategies strictly dominates the other; in the second one, the unique equilibrium involves mixed-strategies (cf. Section 5.3).

In both of these latter contexts, any "monotonic" selection dynamics will tend to make the unique equilibrium configuration stable, provided that the population is large enough.[18] This is obvious in the case of a dominant strategy

---

[16] In the *shifting balance* theory of Wright's (see Sigmund (1993) for a good summary), these small-population effects take place when some small fraction of the whole (large) population happens to be temporarily isolated. In these circumstances, some rare trait may become fixed due to mere chance (something that could hardly occur in very large populations), thus becoming instrumental in the unfolding of a major population shift. Even though our framework is not rich enough to accommodate such ideas fully, the scenario where matching noise remains at significant levels (cf. Theorem 19) reflects similar considerations. Namely the idea that, in not too large populations, the unavoidable noisy interaction which results may render feasible the consolidation of some actions which otherwise would be much less likely to arise.

[17] See Binmore *et al.* (1993) for another instance of an evolutionary framework where long-run outcomes depend on the particular order in which the limits on the different parameters involved is taken.

[18] Rhode and Stegeman (1996) demonstrate that, even for games with a strictly dominant strategy,

since selection-induced adjustment can only proceed in the direction of increasing the number of players choosing this strategy. In the second case, the same *tendency* towards the equilibrium proportions of players is present, but the question becomes slightly more subtle because of the following three complications:

(i) The adjustment may permanently "overshoot" the equilibrium proportions of players if it is not gradual enough. For example, it could repeatedly oscillate between the two monomorphic states if, in every period, adjustment is instantaneous for all players.

(ii) Due to the finiteness of the population, the proportion of it, $\gamma = \frac{d-b}{c-a+d-b}$, which must play strategy $s_1$ in order to mimic the probability weights of the mixed-strategy equilibrium will typically involve a number of players, $\gamma n$, which is *not* an integer.

(iii) Since players adopting strategy $s_1$ do not face the same profile of opponents as those adopting strategy $s_2$ (no player meets himself), the number of players $\rho$ (again, typically not an integer) which would yield an identical pay-off to both strategies (an obvious necessary condition for stationarity of the selection dynamics) is equal to $\rho = \frac{(d-b)\,n+(b-a)}{c-a+d-b}$ (recall Lemma 8). Unless $b = a$, it will *not* be the case that $\gamma n = \rho$.

To remedy (i), let us assume that the selection process is gradual so that, say, only a maximum pre-established number of players may adjust at a time. Further assume, to simplify matters, that we restrict our considerations to "interior" (or polymorphic) states where both strategies are present.[19] Then, if the population is large enough (which, in particular, will decrease matching-induced noise arbitrarily), it is not difficult to see that the selection dynamics will tend to spend "most of its time" in some relatively small set of states whose proportion of $s_1$-adopters is arbitrarily close to $\gamma$ (or $\rho/n$). Due to the considerations explained in (ii) and (iii), one may not generally go beyond such an approximate statement. However, to repeat, the approximation to the mixed-equilibrium profile can be made arbitrarily close if the population is sufficiently large.

Consider now how this state of affairs (for both (5.9) or (5.10)) is affected

the proviso on the size of the population cannot be dispensed with. Specifically, they show that, for *any* given population size, there is a $(2 \times 2)$-game with a dominant strategy for which the unique stochastically stable outcome has the population play the *dominated* strategy. (Notice that the population size is here taken as given, the aim being to find a game where the paradoxical conclusion applies. Thus, the quantifiers on population size and pay-off structure are reversed as compared, say, with the statements of Theorems 19 or 20.) The essential gist of their examples has already been found in Section 2.7 and will again be encountered in the example presented in Section 5.7. In a finite population, a player can always deviate from a certain strategy to become worse off in absolute terms (e.g. if he adopts a dominated strategy) but still make his opponents even worse off.

[19] The essence of the argument is not affected by the fact that the selection dynamics is assumed to leave the monomorphic states fixed. Note that, in the present context, any monomorphic state is destabilized by just one mutation, whereas it will require a large number of mutations (if the population is large) to restore this monomorphic state once the selection dynamics leads the system away from it.

by the introduction of mutation. Since the motion induced by the selection dynamics is always active in the same direction (again, restricting to polymorphic states), mutation can only temporarily offset the tendencies induced by it. Thus, when mutational noise becomes small, the long-run behaviour of the system will follow arbitrarily closely that induced by the selection dynamics. In the long run, therefore, the unique equilibrium is being selected (only approximately for the case given by (5.10)), both when the game has a dominant strategy and when the unique equilibrium involves mixed strategies.

The previous heuristic discussion suggests that the postulated evolutionary approach behaves in a natural ("non-pathological") fashion when equilibrium selection is not an issue, i.e. when either (5.9) or (5.10) applies. A formal proof of this conclusion can be found in Kandori *et al.* (1993) and Robson and Vega-Redondo (1996) for each of the scenarios considered here (i.e. under small or large matching noise, respectively).[20]

## 5.5    Continuous-Time Dynamics

As in much of the recent evolutionary literature, the dynamic processes studied in this chapter have been modelled in a discrete temporal framework. There are, however, a few important exceptions where time has been modelled continuously. Two such instances are now briefly summarized.

One of them, the seminal work of Foster and Young (1990), represents the first attempt at explicitly modelling ceaseless stochastic noise in evolutionary contexts. Not only did it provide much of the conceptual thrust which fuelled the early developments of this literature, but it also imported from the mathematical arena some of the powerful analytical tools which have been extensively used since (specifically, the techniques developed by Freidlin and Wentzel (1984), used in the proofs of Theorems 19 and 20). These authors focused on the standard (continuous-time) Replicator Dynamics (3.4), adding to it a (variably scaled) Wiener process, i.e. a continuous white noise process with mean zero and unit variance-covariance matrix. Specifically, that is, they studied a stochastic dynamical system on $\Delta^{m-1}$ (the $(m-1)$-dimensional simplex) of the following form:[21]

$$d\nu_i(t) = \nu_i(t) \left\{ [A_i \nu(t) - \nu(t) \cdot A \nu(t)] \, dt + \sigma \left[ \Gamma_i(\nu(t)) \, dW(t) \right] \right\}, \quad (5.20)$$

---

[20] Other extensions for the scenario with small matching noise are discussed by Kandori and Rob (1995) who address strict co-ordination games (games where all off-diagonal pay-offs are zero) and general supermodular games (see Topkis (1979)). Robson and Vega-Redondo (1996) also explore some generalizations of Theorem 19. In particular, they show that efficient selection still applies to *any* common-interest game (cf. Definition 4), even if the game is not symmetric and two different populations are involved.

[21] The notation here is as in Ch. 3. The pay-off matrix is denoted by $A$, with $A_i$ standing for its $i$th row. The vector of population frequencies is denoted by $\nu(t) = (\nu_1(t), ..., \nu_m(t))$.

for all $i = 1, 2, ..., m$, where $\sigma \in \Re_+$ is a parameter scaling the magnitude of noise, $W(t)$ is the described $m$-dimensional Wiener process, and $\Gamma(\nu) \in \Re^{m \times m}$ is continuous in $\nu$, bounded away from zero, and has the property that $\nu \cdot \Gamma(\nu) \equiv (0, 0, ..., 0)$, for all $\nu \in \Delta^{m-1}$.

The noise term added in (5.20) represents a "catch-all" *aggregate* disturbance which can be provided with a number of different interpretations. One possibility is to view it as capturing the unavoidable uncertainties which arise when a finite-population context (as, of course, all of them are in the real world) is approximated by a continuum-population model. Another possible interpretation is that any given population (independently of its size) confronts sources of uncertainty (e.g. exogenous changes in the environment) which will affect all of them in a correlated fashion and, therefore, cannot be ignored by resorting to large-number arguments. In any case, even if these effects are small, the basic tenet of the approach pioneered by Foster and Young is that they may well have very significant qualitative effects on the long-run behaviour of the system.

Under the implicit assumption that the system is subject to some background process of mutation (or migration) by which all strategies remain represented in the population at some significant frequencies, Foster and Young (1990) assume that the analysis of the system can be restricted to some subset $S_\delta = \{\nu \in \Delta^{m-1} : \nu_i \geq \delta\} \subset \Delta^{m-1}$ where $\delta > 0$ is chosen sufficiently small. This assumption has been incorporated explicitly into the theoretical framework of Fudenberg and Harris (1992), as explained below.

Foster and Young pose the following, by now familiar-sounding, question: What is the long-run behaviour of the system as $\sigma$, the noise parameter, converges to zero? First, they show that this question can always be provided with a well-defined and definite answer (at least, focusing on the limit superior, as $\sigma \to 0$). Second, they illustrate by means of examples that this approach is a fruitful one and provides interesting insights.

In particular, they show that, in some contexts where several ESS configurations exist, their approach is able to single out one of them as the unique *stochastically stable state*, i.e. the unique state which retains positive density in the limit ergodic distribution. On the other hand, they also show that there are scenarios where a unique ESS exists but whose set of stochastically stable states is completely disjoint from the support of the ESS population profile. Thus, in this latter sense, one arrives at the following conclusion: Even though the ESS concept may embody a suitable notion of dynamic stability when the environment displays very rare perturbations (cf. Theorem 3), it represents a much less appropriate concept in case of ceaseless perturbations.

This approach has been developed further by Fudenberg and Harris (1992), who suggest that the best way to introduce noise into the Replicator Dynamics is by incorporating it directly in the underlying dynamics in population *sizes* (rather than frequencies) which is taken to generate it (recall Section 3.2). They argue that, in this fashion, one is better able to discern the appropriate natural restrictions to be imposed on the stochastic component of the dynamics. Since, in a sense, the Replicator Dynamics is a reduced model, any assumptions directly made on it may hide unsuitable requirements on the underlying, more "primitive" model.

Denote by $r(t) = (r_1(t), r_2(t), ..., r_m(t)) \in \Re^m$ the vector of population sizes expressing the "number" of individuals adopting each of the $m$ strategies at $t$, and let $\nu(t) = (\nu_1(t), ..., \nu_m(t))$ be the associated population frequencies, i.e.

$$\nu_i(t) \equiv \frac{r_i(t)}{\sum_{j=1}^m r_j(t)}, \quad i = 1, 2, ..., m.$$

Fudenberg and Harris propose the following stochastic dynamical system:

$$dr_i(t) = r_i(t) \{[A_i \nu(t) - \nu(t) \cdot A \nu(t)] \, dt + \sigma_i \, dW_i(t)\}, \quad i = 1, 2, ..., m,$$
(5.21)

where again $W(t) = (W_1(t), ..., W_m(t))$ is an $m$-dimensional Wiener process and $\sigma = (\sigma_1, ..., \sigma_m) \in \Re^m$ is a vector of parameters scaling noise. Applying Ito's Lemma (see Gihman and Skorohod (1972)), one may derive the induced dynamics for population frequencies. For example, in the simple context with just two strategies $(m = 2)$, it may be represented as follows:[22]

$$d\nu_1(t) = \nu_1(t) \, \nu_2(t) \left\{ \begin{array}{c} [A_1 \nu(t) - A_2 \nu(t)] \, dt + [\nu_1(t) \, (\sigma_1)^2 - \nu_2(t) \, (\sigma_2)^2] \, dt \\ +\sigma_1 \, dW_1(t) - \sigma_2 \, dW_2(t) \end{array} \right\}.$$
(5.22)

Suppose that the pay-off matrix $A$ included in (5.22) defines a co-ordination game. (That is, a game as in Table 15 with $a > c$, $b > d$.) In that case, Fudenberg and Harris show that the system will eventually become *absorbed* by an essentially monomorphic profile with probability one. However, with positive $\sigma_1$ and $\sigma_2$, the stochastic nature of the system makes the prediction of which of the two monomorphic situations will prevail (i.e. one with $\nu_1 = 1$ or $\nu_2 = 1$) far from certain. *Ex ante*, the only statement that can be made is on the relative *prior* probabilities of achieving each of them. Naturally, these probabilities will depend on the initial conditions, relative basins of attraction, and noise parameters. For example, in the limit case where $\sigma_1$ and $\sigma_2$ converge to zero, initial conditions and basins of attraction become pre-eminent. And then it can be asserted that the situation achieved in the long run will be, with arbitrarily high probability, the one corresponding to the basin of attraction where the initial conditions happen to lie.

If, instead, the pay-off matrix $A$ defines a game with either a dominant strategy or a unique mixed-strategy equilibrium – the other two possible (generic) $(2 \times 2)$-scenarios – the process is always ergodic. Furthermore, its long-run probability is fully concentrated on the population state which mimics the unique equilibrium.

---

[22] Since the system is one-dimensional, only the law of motion for one of the strategies (say the first one) needs to be specified.

To regain ergodicity in the co-ordination case, Fudenberg and Harris modify
(5.22) by adding a constant flow of mutation. Generalizing the formulation described in Sections 3.9 and 4.8 (where the perturbation was channelled through
unbiased newcomers replacing incumbents at a uniform rate) mutation is identified here with a given conversion flow across populations playing different
strategies.[23] Specifically, if $\lambda_i$ $(i = 1, 2)$ denotes the rate at which individuals
adopting strategy $s_i$ switch to strategy $s_j$ $(j = 3 - i)$, the system (5.22) is
modified as follows:

$$dr_i(t) = r_i(t) \left\{ [A_i \nu(t) - \nu(t) \cdot A \nu(t)] \, dt + \sigma_i \, dW_i(t) \right\} - \lambda_i r_i(t) + \lambda_j r_j(t),$$
(5.23)

for each $i, j = 1, 2, j = 3 - i$. Applying again Ito's Lemma to derive a corresponding system in population frequencies, one obtains:

$$d\nu_1(t) = \nu_1(t) \, \nu_2(t) \left\{ \begin{array}{c} [A_1 \nu(t) - A_2 \nu(t)] \, dt \\ + [\nu_1(t) \, (\sigma_1)^2 - \nu_2(t) \, (\sigma_2)^2] \, dt + (\lambda_2 - \lambda_1) \, dt \\ + \sigma_1 \, dW_1(t) - \sigma_2 \, dW_2(t) \end{array} \right\}$$
$$+ \left[ \lambda_2(\nu_2(t))^2 - \lambda_1(\nu_1(t))^2 \right] \, dt.$$
(5.24)

In this context, it is shown that as every $\sigma_i$ and $\lambda_i$ converge to zero (in any
order, provided the ratio $\lambda_1/\lambda_2$ remains constant) the process remains ergodic,
the limit long-run distribution of the process concentrating its full mass in the
risk-dominant equilibrium.

It is important to understand the intuition underlying the sharp contrast between the long-run behaviour of (5.22) and (5.24). This contrast underscores
the qualitative differences between pay-off-perturbing noise – i.e. that induced
by the Wiener process $W(t)$ – and mutation. The first one always maintains
a large potential influence throughout the state space in affecting the *direction*
of movement of the system. However, as the process approaches the boundary,
the *magnitude* of the effect vanishes as the factor $\nu_1(t) \, \nu_2(t)$ appearing in (5.24)
becomes very small. It is precisely these considerations which cause the process
to be non-ergodic in the absence of mutation: eventually, the process will become "trapped" near one of the boundaries, pay-off-perturbing noise becoming
progressively more unlikely to bring it out of this situation.

To prevent such a state of affairs is precisely the function of the mutation
rates in (5.24). No matter how small they are, they always have the effect of
keeping the system within some compact subset of $int(\Delta^1)$, even though their
*absolute* impact in "bending" the system away from its boundaries might be
very small. In these circumstances, pay-off-perturbing noise is sure to maintain
its role in having the system escape the basin of attraction of either equilibrium.
And again, given such a guaranteed role for (small) noise, the equilibrium that

[23] Note that, if $\lambda_1 = \lambda_2 = \lambda$, this formulation is essentially identical to that postulated in
Sections 3.9 and 4.8. In this case, the mutation term in (5.23) below becomes $\lambda(r_j(t) - r_i(t)) =$
$2\lambda(\frac{1}{2} - r_i(t))$, which is formally analogous to the one postulated there.

will be selected in the long-run turns out to be the risk-dominant one, due to a similar kind of reasoning to that which underlies Theorem 20. (Note that the present framework displays no matching noise, due to the assumption of large numbers implicitly built in the Replicator Dynamics.)

Given that similar considerations seem to arise from both the continuous- and discrete-time frameworks, one may wonder whether they can be conceived as essentially equivalent. Even though much less work has been carried out in the continuous-time framework, an important distinction between the two ap- proaches lies in their differing *robustness* to alternative particular specifications. As described above, the discrete-time approach merely demands that some gen- eral (qualitative) requirements of monotonicity be met, thus abstracting from any precise formulation of the selection dynamics. This, however, is not the case for the evolutionary framework modelled in continuous time. In this latter context, the analysis will, in general, be highly dependent on the particular formulation adopted. For example, the above-described conclusions will *not* carry over to other monotonic selection dynamics different from the Replicator Dynamics. In particular, one may find perturbed monotonic dynamics whose limit distribution is concentrated on the risk-dominated equilibrium.

One must understand, however, that the discrete-time approach builds its robustness upon the following critical and potentially controversial feature: un- der small matching noise, transition probabilities across different equilibria will only be dependent on the respective *sizes* of their basins of attraction. This arises from the fact that, typically, such transitions will be conducted through large "jumps". (In the continuous-time framework, what is needed is a "lucky *concatenation*" of suitable random events.) Since such large jumps require a correspondingly large number of *simultaneous* mutations, their likelihood be- comes very small as the mutation probability also becomes small. This raises the question of how one should interpret long-run results which, if the tran- sition probabilities are very small, may take an unduly large amount of time to materialize. A discussion of these issues, together with their relationship to some important details of the theoretical framework (specifically, the postulated interaction pattern), is the object of the next section.

## 5.6   Rate of Convergence and Interaction Pattern

### *5.6.1    Global interaction*

Theorems 19 and 20 have been interpreted as statements of equilibrium selection in games, even though they are couched in a somewhat unusual form. In particular, they do *not* assert that the system will eventually converge to a certain outcome (something precluded from the recurrent operation of mutation) but, rather, that a.s. the system will spend "most of its time" along any sample path at some specific state.

These results do not convey in themselves any idea on how long the long run might be. Or, to express it somewhat more precisely, they do not provide any information on whether one should expect the stated long-run conclusion to be realized at a fast or slow rate. As suggested above, it might be rightly argued that if the expected rate at which the system achieves its long-run outcome is very slow, the independence of initial conditions that these results formally display is only of limited relevance as an equilibrium selection device.

Clearly, any assessment on the rate at which the long-run state of affairs is to be realized must depend, very crucially, on the mutation rate $\epsilon$ under consideration. Mutation is the phenomenon which essentially carries the "ergodicity burden" in the model. It is mutation which eventually makes initial conditions irrelevant by, sooner or later, upsetting the forces of selection. It is apparent, therefore, that any precise assessment of the rate of long-run convergence must be related to the probabilities that certain mutation events (typically a combination of them) happen to take place. In the end, as $\epsilon$ becomes small, these probabilities must be of the same order as some appropriate power of $\epsilon$, the mutation rate.

The speed at which the process converges to its long-run state of affairs may be formally evaluated in a variety of different ways, all of them basically equivalent. For example, in contexts such as the present one where the (limit) invariant distribution is known to be fully concentrated in a single long-run (stochastically stable) state, the question can be posed in either one of the following ways:

(a) Suppose the system is not in the long-run state. What is the maximum *expected* time that will elapse before the process visits this state?

(b) Choose the *initial* state of the system with some arbitrary prior probability. What is the fastest rate (exponential on $t$)[24] at which the probability of being in the long-run state at some future $t$ converges to one?

Only the approach suggested by (a) will be formally discussed here. The reader can find in Ellison (1993) a detailed discussion of approach (b) for a context essentially analogous to one with small matching noise (i.e. one where players form static expectations – see Chapter 6). As mentioned, both (a) and (b) yield the same insights.

First, the focus is on how the rate of convergence is affected by the different matching-noise scenarios contemplated above. As established in Subsections 5.4.1 and 5.4.2, the long-run outcome is generally different in each of these scenarios. Here, it is shown that they also display drastically different rates of long-run convergence.

---

[24] As is well known from standard results in the Theory on Stochastic Processes (see e.g. Karlin and Taylor (1975)), convergence towards the invariant distribution is always at an exponential rate.

To simplify matters, the analysis will focus on the especially simple specification of the selection dynamics in which adjustment is assumed instantaneous.[25] Thus (as postulated in Section 5.2 too) the selection component of the evolutionary process is assumed to be as follows:

$$B(z_t, p_t^v) = \left\{ \begin{array}{ll} n & \text{if } \pi_1(z_t, p_t^v) > \pi_2(z_t, p_t^v) \\ 0 & \text{if } \pi_1(z_t, p_t^v) < \pi_2(z_t, p_t^v) . \end{array} \right. \tag{5.25}$$

Relying on previous notation, it becomes

$$\hat{B}(z_t) = \left\{ \begin{array}{ll} n & \text{if } \hat{\pi}_1(z_t) > \hat{\pi}_2(z_t) \\ 0 & \text{if } \hat{\pi}_1(z_t) < \hat{\pi}_2(z_t) \end{array} \right. \tag{5.26}$$

when matching noise is made to vanish by making the number of rounds grow unboundedly.

The contrast between the two scenarios is then established by the following two respective Propositions.

**Proposition 12** *Under the conditions specified in Theorem 19, suppose that the selection dynamics is given by (5.25) for any given $v \in \mathcal{N}$. $\exists \alpha > 0$ such that, for small enough $\epsilon > 0$, the maximum expected time $\tau^*$ at which the process first visits the unique long-run state $\omega^1 = (n, 0)$ satisfies $\tau^* \leq \alpha \epsilon^{-2}$.*

**Proposition 13** *Under the conditions specified in Theorem 20, suppose that the selection dynamics is given by (5.26). Let $\gamma = \frac{d-b}{c-a+d-b}$ and denote $\Delta \equiv |a - b|$. $\exists \beta > 0$ such that, for small enough $\epsilon > 0$, the maximum expected time $\hat{\tau}$ at which the process first visits the unique long-run state $\omega^2 = (0, n)$ satisfies $\hat{\tau} \geq \beta \epsilon^{-\gamma n + \Delta}$.*

**Proof of Proposition 12.**   Given any $\epsilon > 0$, consider an evolutionary process as described, with the only exception that it makes the state $\omega^1$ absorbing. Denote by $Q_\epsilon(\cdot, \cdot)$ its corresponding transition matrix. The proof of the result easily follows from the following claim.

*Claim 3:* Let $\omega = (z, n - z) \in \Omega$ be any arbitrary state of the process. There is some $\eta > 0$, independent of $\epsilon$ and $\omega$, such that $Q_\epsilon^{(2)}(\omega, \omega^1) \geq \eta \epsilon^2$.

To prove this claim, let $\omega' = (z', n - z')$ be any state for which $Q_0(\omega, \omega') > 0$, i.e. a transition from $\omega$ may occur without resorting to mutation. Denote by $z''$ the least positive even number such that $z'' \geq z'$. Obviously, $z'' - z' \leq 2$, which implies that at most two mutations will transform state $\omega'$ into a state where there is some positive and even number of players adopting strategy $s_1$. Given that some such state has been reached, assume that all $s_1$-adopters are matched among themselves in each of the $v$ matching rounds. In this event,

[25] The gist of the conclusions, however, is fully general, as can be found formally proved in Ellison (1993) or Robson and Vega-Redondo (1996).

the ensuing transition must be towards state $\omega^1$ with probability one, in view of (5.25). This implies that, in at most two periods, the process with transition matrix $Q_\epsilon$ will be absorbed at $\omega^1$ with a probability that, for $\epsilon$ small, is of an order no lower than $\epsilon^2$. That is, it is bounded below by some $\eta_\omega \epsilon^2$ for some $\eta_\omega > 0$ independent of $\epsilon$. The proof of the claim is then completed by making $\eta = \min_{\omega \in \Omega} \{\eta_\omega\}$, which is positive due to the finiteness of the state space.

Claim 3 implies that the maximum expected time to absorption of the process with transition matrix $Q_\epsilon$ (an upper bound for $\tau^*$) must itself be bounded above by twice that of a Poisson process with arrival rate $\eta \epsilon^2$. That is, it must be bounded above by $2(\eta \epsilon^2)^{-1}$. Choosing $\alpha = 2\eta^{-1}$, the proof of the Proposition is complete. ■

**Proof of Proposition 13.** To find a lower bound on the maximum expected time of transition to state $\omega^2$, consider specifically that corresponding to the transition from $\omega^1$. In view of (5.26) and the considerations explained in the proof of Theorem 20, this transition will require a number of mutations no smaller than $\gamma n - \Delta$. (Fewer mutations from state $\omega^1$ will immediately return the process to this state next period.) For small $\epsilon$, the probability of this event is of order no higher than $\epsilon^{\gamma n - \Delta}$. That is, there is some $\zeta > 0$ such that this probability is bounded above by $\zeta \epsilon^{\gamma n - \Delta}$ for sufficiently small $\epsilon$. Consequently, the expected time for this event to materialize (again, bounding it by that of a corresponding Poisson process) is $(\zeta \epsilon^{\gamma n - \Delta})^{-1}$. Choosing $\beta = \zeta^{-1}$, the proof of the Proposition is complete. ■

Contrasting Propositions 12 and 13, one observes a qualitatively very significant difference between the speeds of convergence prevailing in each case. Whereas in the context of Proposition 12, the order in $\epsilon$ of the maximum expected waiting time does not depend on population size,[26] it does so in an affine increasing way in the context of Proposition 13. If the population is large (but given),[27] this amounts to a substantial difference in the rate of long-run convergence between both cases. This formalizes in a precise way the heuristic role formerly attributed to matching noise as a potential "amplifier" (a biased one, however) of mutational noise – cf. Subsection 5.4.3.

---

[26] In the general case where selection adjustment is not necessarily instantaneous, this expected waiting time may be of order lower than $\epsilon^{-2}$, but it still is independent of population size.

[27] Of course, if one allows population size to grow, this will have an effect on matching-induced noise analogous to that of increasing the number of rounds. In that case, again one would have to compare the different rates at which population grows (with its corresponding effect on matching noise) and the rate at which the mutation probability becomes small.

### 5.6.2    *Local interaction*

The role played by population size in the context of Proposition 13 (with small matching noise) seems inherently associated with the fact that interaction and selection are postulated to be fully global phenomena. In most real-world contexts, however, it seems more appropriate to conceive both of these components of the model as having quite a local dimension. Agents, that is, usually tend to interact more often with a certain subset of neighbours, also tailoring their action adjustments to the pay-offs and actions these neighbours display.

This, to be sure, need not eliminate matching uncertainty. For any given agent, it will simply be narrowed down to his respective set of "neighbours". Thus, local interaction is still conceptually consistent with the presence of significant matching noise. And if this is the case, the analysis on the long-run performance of the model which was conducted in Subsection 5.4.1 remains essentially unaffected. Even the conclusion of Proposition 12 on relatively fast (population-independent) rate of long-run convergence applies without any significant variation.

In contrast, one would expect that the analysis conducted for small matching noise would be significantly affected by the consideration of a local interaction pattern. As suggested above, at least the result on the rate of convergence established by Proposition 13 appears to be crucially dependent on the postulated pattern of global interaction.

To explore this issue, a very simple framework of local interaction is now proposed, similar to that of Ellison (1993). Within this model, the previous intuitive conjecture is confirmed, i.e. the rate of long-run convergence is substantially faster. On the other hand, the nature of the long-run outcome is also affected since, as explained below, the force of risk-dominance as a selection criterion becomes somewhat weakened.

Assume that the population is arranged in a "circle", with each individual $i = 1, 2, ..., n$ having exactly two neighbours: agent $[i - 1]$ and agent $[i + 1]$, where $[\cdot]$ denotes "modulo $n$". (That is, $[i] = i$ for $i = 1, 2, ..., n$, but $[0] = n$ and $[n + 1] = 1$.) Every individual meets each one of his two neighbours some arbitrarily given number of times (the same number of times for each neighbour)[28] in order to play a game as described in Table 15.

In this context, the state space cannot be formulated any longer in an anonymous way. Now, the state of the system must specify the strategy adopted by each (well-identified) player. Thus, the state space is made equal to $\Theta = \{s_1, s_2\}^n$ where each $\theta \in \Theta$ is the $n$-tuple of current strategies adopted by each player.

Let us now postulate a selection dynamics that is formulated in the spirit of

---

[28] This is assumed in order to remove all matching uncertainty in the simplest manner, thus allowing us to contrast the present analysis with that carried out in Subsection 5.4.2. This assumption could again be rationalized by a suitable application of the Law of Large Numbers if the number of matching rounds is taken to be large enough and the probability that each individual meets either of his neighbours is identical.

(5.26) but reflects the local nature of the interaction. Denote by $\hat{B}^i(\theta_t)$ the new strategy adopted by player $i$ at the selection phase in $t + 1$ when the previous state is $\theta_t$. Furthermore, let $\hat{\pi}_1^i(\theta_t)$ and $\hat{\pi}_2^i(\theta_t)$ stand for the respective average pay-offs earned by strategies $s_1$ and $s_2$ in the neighbourhood of player $i$ (that is, by players $[i-1]$, $[i]$ and $[i+1]$), with the convention that they equal $-\infty$ if the corresponding strategy is not adopted in this neighbourhood.[29] The selection component of the dynamics is simply formulated as follows:

$$\hat{B}^i(\theta_t) = \left\{ \begin{array}{ll} s_1 & \text{if } \hat{\pi}_1^i(\theta_t) > \hat{\pi}_2^i(\theta_t) \\ s_2 & \text{if } \hat{\pi}_1^i(\theta_t) < \hat{\pi}_2^i(\theta_t) \end{array} \right. \tag{5.27}$$

for each $i = 1, 2, ..., n$.

As in the original model, assume that, every period, each player mutates with an independent probability $\epsilon$ just before entering any round of matching and play. This renders the evolutionary process ergodic, its unique invariant distribution being denoted by $\lambda_\epsilon(\cdot) \in \Delta(\Theta)$. Again, the analysis will focus on the limit invariant distribution obtained when the mutation probability $\epsilon$ becomes arbitrarily small.

As advanced, not only the rate of convergence but also the long-run outcome is affected by the local nature of the interaction. As before, there is a tendency for risk-dominance to prevail over efficiency, although in a weaker form. Now, the inefficient equilibrium will only be selected in the long run when the efficient one is substantially more "risky" than before. Specifically, it turns out that in order for $\theta^2 \equiv (s_2, s_2, ..., s_2)$ to be the unique stochastically stable state the following inequality (in terms of the pay-offs specified in Table 15) must hold:

$$2(b + c) > 3a + d. \tag{5.28}$$

It is easy to check that this inequality implies that $b + c > a + d$, i.e. the strategy $s_2$ is risk-dominant. In general, of course, (5.28) is consistent with $s_1$ being the efficient strategy, i.e. $a > b$.

A combined statement on both the long-run outcome and the associated rate of long-run convergence for the present context is contained in the following result.

**Theorem 21** *Assume (5.8), (5.28), $n \geq 5$, and let $\theta^2$ be the monomorphic state where all players adopt $s_2$. Then:*
*(a) $\lim_{\epsilon \to 0} \lambda_\epsilon(\theta^2) = 1$.*
*(b) $\exists \chi > 0$ such that, for small enough $\epsilon > 0$, the maximum expected time $\tilde{\tau}$ at which the process first visits the (unique) long-run state $\theta^2$ satisfies $\tilde{\tau} \leq \chi \epsilon^{-2}$.*

**Proof.** Since it relies on arguments very analogous to those used in previous analysis, the proof will be merely sketched. To show (a), the main observa-

---

[29] Note that, as was also generally the case in (5.11), the average pay-offs for each strategy are not computed over the same number of occurrences. In particular, it can never be the case in the present context that the two strategies are played the same number of times in any of the three-individual neighbourhoods.

tion to make is that, as long as there is one connected string of (at least two) neighbouring individuals choosing $s_2$, (5.27) and (5.28) imply:

*Claim 4:* Every $s_1$-adopter $i$ who is *adjacent* to such an $s_2$-string will adjust his strategy towards $s_2$.

This claim follows from the following straightforward computations. In the neighbourhood of $i$, the average pay-off to strategy $s_1$, $\hat{\pi}_1^i$, is bounded above by $\left[\frac{1}{2}(2a) + \frac{1}{2}(a+d)\right]$ if one of his neighbours also does $s_1$ (which will be labelled Case 1) or simply equals $2d$ otherwise (Case 2). On the other hand, the average pay-off to strategy $s_2$ in the neighbourhood of $i$, $\hat{\pi}_2^i$, is either $(b+c)$ in Case 1, or bounded below by $\left[\frac{1}{2}(b+c) + \frac{1}{2}(\min\{b,c\}+c)\right]$ in Case 2. Under (5.28), it can easily be verified that $\hat{\pi}_2^i > \hat{\pi}_1^i$ in either Case 1 or Case 2.

*Claim 5:* Every $s_2$-adopter $j$ who is in the frontier of the $s_2$-string (in particular, has neighbours doing different strategies on each side) will stay with strategy $s_2$.

This claim follows from the following two facts. On the one hand, $\hat{\pi}_1^j$ is bounded above by $a+d$, the maximum pay-off which can be achieved by $j$'s only neighbour choosing $s_1$. On the other hand, $\hat{\pi}_2^j$ is bounded below by $\left[\frac{1}{2}(b+c) + \frac{1}{2}(b+\min\{b,c\})\right]$, where the actual value achieved by this expression depends on whether individual $j$ forms part of an $s_2$-string with two or more individuals. In any case, again (5.28) implies $\hat{\pi}_2^j > \hat{\pi}_1^j$.

The previous two claims indicate that just two *adjacent* mutations can trigger a transition from $\theta^1 = (s_1, s_1, ..., s_1)$ to $\theta^2 = (s_2, s_2, ..., s_2)$. However, the converse transition will require more than two mutations if $n \geq 5$. (In this case, only two mutations from $\theta^2$ will always leave a string of at least two individuals playing $s_2$.) Proceeding along the lines used in previous proofs, one can rely on these two insights to prove that, for small $\epsilon > 0$, the unique long-run outcome of the process is indeed $\theta^2$.

Finally, the fact that a two-individual string of $s_2$-adopters is by itself enough to cause a transition to state $\theta^2$ guarantees (by an argument fully analogous to that used in Proposition 12) that the maximum expected waiting time for visiting the long-run state $\theta^2$ is of the order $\epsilon^{-2}$, for small $\epsilon$. This completes the proof. ∎

In contrast with Theorem 20, the previous result confirms the intuition that a local interaction pattern will increase substantially the rate of long-run convergence. In particular, Theorem 21 shows that this rate becomes independent of population size, an upper bound on it being established which is of the same order as that established by Proposition 12 for the large matching-noise scenario.

The previous result also demands a more stringent requirement of risk domination for an inefficient outcome to be selected in the long run. (Note, of course, that the violation of (5.28) is consistent with the usual risk-dominance criterion reflected by the inequality $b + c > a + d$.) In fact, it is not difficult to show that, in the local-interaction context considered here, the efficient outcome may be selected in the long run even though the alternative equilibrium is risk-dominant.

## 5.7  The Evolution of Walrasian Behaviour

We close this chapter with a simple evolutionary model of Cournotian oligopoly, taken from Vega-Redondo (1996c).[30] The purpose of this application is twofold. First, it illustrates the wide applicability of evolutionary analysis to economic contexts, well beyond the stylized random-matching scenarios used above to introduce the basic ideas. Second, it shows that evolutionary processes may represent much more than mere equilibrium-selection devices, and have the potential of inducing interesting non-Nash behaviour.[31]

The context is as described in Subsection 2.7.2. A set of firms $N = \{1, 2, ..., n\}$ is involved in a market for an homogeneous product with an inverse-demand function $P : \Re_+ \rightarrow \Re_+$, assumed decreasing (i.e. satisfying the Law of Demand). Every firm $i = 1, 2, ..., n$ has an identical cost function $C : \Re_+ \rightarrow \Re_+$ which determines its cost $C(x_i)$ of producing each possible output $x_i$. No particular condition is demanded from this cost function other than it should allow for the existence of a symmetric Walrasian equilibrium. (Thus, for example, it cannot display decreasing marginal costs throughout.)

A symmetric Walrasian equilibrium (cf. (2.19)) is defined as a monomorphic configuration where the output $x^w$ produced by *each* firm satisfies:

$$P(n\,x^w)\,x^w - C(x^w) \geq P(n\,x^w)\,x - C(x), \qquad (5.29)$$

that is, maximizes its profits, taking the induced market-clearing price as given. (The argument used in the proof below guarantees that, provided it exists, a symmetric Walrasian equilibrium is unique.)

In order to remain within the context of a finite-state Markov chain, it is assumed that firms have to choose their output from a common real grid $\Gamma = \{0, \delta, 2\delta, ..., r\delta\}$ for any arbitrary $\delta > 0$ and $r \in \mathcal{N}$. The only requirement is that $x^w \in \Gamma$.

As time proceeds, firms are postulated to adjust their output by simply mimicking that of a firm which is earning (i.e. has earned the previous period) the highest profit. Occasionally, they also mutate. To define this combined dynamics precisely, some notation is first introduced.

---

[30] Rhode and Stegeman (1995) develop a similar application focused on a two-firm (i.e. duopoly) scenario.

[31] In a sense, what is done here can be seen as a dynamic elaboration of the analysis carried out in Subsection 2.7.2 on the "evolutionary stability" of Walrasian behaviour. In contrast with our former static analysis, this behaviour is now shown to be *uniquely* evolutionarily stable (in the stochastic sense).

At each $t = 0, 1, 2, ...$, the state of the system $\omega_t$ is identified with the current output profile $(x_{1t}, x_{2t}, ..., x_{nt})$. Associated with it, the corresponding vector of profits $\pi_t = (\pi_{1t}, \pi_{2t}, ..., \pi_{nt})$ is given by

$$\pi_{it} = P(\sum_{i=1}^{n} x_{it}) \, x_{it} - C(x_{it}), \quad i = 1, 2, ..., n.$$

On the basis of these realized profits, the set of "best outputs" at $t$ is defined as

$$B_t = \{x_{it} : \pi_{it} \geq \pi_{jt}, \, \forall j = 1, 2, ..., n\}.$$

Selection dynamics is formulated as follows. At every $t$, each firm $i = 1, 2, ..., n$ enjoys a common independent probability $p \in (0, 1)$ of revising its former output. In this event, it is assumed to choose any output from the set $B_{t-1}$, i.e. the previous set of best outputs, according to a certain probability distribution with full support.[32]

On the other hand, mutation is formalized as usual. At every $t$, and once the selection adjustment has been completed, every firm changes its output with some probability $\epsilon > 0$, the new output being selected according to some given probability distribution with full support on $\Gamma$. After selection and mutation have been completed, pay-offs materialize, which then leads into the ensuing period.

Given $\epsilon > 0$, let $\mu_\epsilon(\cdot)$ denote the unique invariant distribution associated with the corresponding ergodic process. The following result establishes a clear-cut identity between its (unique) stochastically stable outcome and Walrasian behaviour.

**Proposition 14 (Vega-Redondo (1996c))** *Let $\hat{\omega} = (x^w, x^w, ..., x^w)$. Then,* $\lim_{\epsilon \to 0} \mu_\epsilon(\hat{\omega}) = 1$.

**Proof.** The argument relies on the graph-theoretic techniques of Freidlin and Wentzel (1984) used throughout this chapter. The previous concepts and notation are readily adapted to the present context. On this basis, the proof of the Proposition follows from the following three Lemmas and the usual complementary arguments.

**Lemma 11** *Some set $A \subset \Omega$ is a limit set of the mutation-free dynamics if, and only if, it is a singleton $\{\omega\}$ consisting of a monomorphic state $\omega$ (i.e. $\omega = (q, q, ..., q)$ for some common $q \in \Gamma$).*

**Lemma 12** *There exists an $\hat{\omega}$-tree $\hat{Y} \in \mathcal{Y}_{\hat{\omega}}$ such that $c(\hat{Y}) = r$, where recall that $r + 1$ is the cardinality of the grid $\Gamma$.*

**Lemma 13** *For all monomorphic states $\tilde{\omega} \neq \hat{\omega}$, any $\tilde{\omega}$-tree $\tilde{Y} \in \mathcal{Y}_{\tilde{\omega}}$ has $c(\tilde{Y}) \geq r + 1$.*

---

[32] As explained in Vega-Redondo (1996c), this simple formulation can be substantially generalized.

*Proof of Lemma 11*   The sufficiency part is obvious from the specification of the process (i.e. if a state is monomorphic, no strategy revision possibilities will alter the state). On the other hand, the necessity follows from the fact that strategy revision is a firm-independent phenomenon whose probability density at each $t$ is assumed to have full support on the (common) set $B_{t-1}$. Therefore, there is always positive probability (bounded above zero, since the state space is finite) that all firms adjust their strategy towards the same output.

*Proof of Lemma 12.*   Let $\omega_x = (x, ..., x)$ be any monomorphic (limit) state with $x \neq x^w$. It is first claimed that there exists a path $y_x$ linking $\omega_x$ to $\hat{\omega}$ such that $c(y_x) = 1$.

Let $\omega'_x$ be a state where all firms choose output $x$ except for a single (arbitrary) firm which chooses the Walrasian output $x^w$. Make $y_x = \{(\omega_x, \omega'_x), (\omega'_x, \hat{\omega})\}$. Obviously, $c(\omega_x, \omega'_x) = 1$. To establish now that $c(\omega'_x, \hat{\omega}) = 0$, it is enough to show that in state $\omega'_x$ the firm that produces $x^w$ obtains strictly higher profits than the rest. That is,

$$P((n-1)x + x^w) x^w - C(x^w) > P((n-1)x + x^w) x - C(x). \qquad (5.30)$$

The argument is quite similar to that developed in Subsection 2.7.2. Since $P(\cdot)$ is strictly decreasing, one has:

$$[P(nx^w) - P((n-1)x + x^w)] x^w < [P(nx^w) - P((n-1)x + x^w)] x.$$

Or equivalently,

$$P(nx^w) x^w + P((n-1)x + x^w) x < P(nx^w) x + P((n-1)x + x^w) x^w.$$

Subtracting the term $[C(x) + C(x^w)]$ from both sides of the above expression it follows that

$$\begin{aligned} [P(nx^w) x^w - C(x^w)] + [P((n-1)x + x^w) x - C(x)] < \\ [P(nx^w) x - C(x)] + [P((n-1)x + x^w) x^w - C(x^w)]. \end{aligned} \qquad (5.31)$$

From (5.29), the first term in the LHS of (5.31) is no smaller than the first term in its RHS. Therefore, it must be the case that the second term in the LHS of this expression is strictly smaller than its second term in the RHS. But this is just what (5.30) expresses, which confirms the desired claim.

It is now verified that there is a $\hat{\omega}$-tree $\hat{Y}$ whose cost $c(\hat{Y}) = r$. To construct such a tree consider first the $2r$ links $\{(\omega_x, \omega'_x), (\omega'_x, \hat{\omega}) : x \neq x^w\}$. The aggregate cost of these links is $r$. But then, since monomorphic states are the only limit sets of the mutation-free dynamics (recall above), the remaining states can be linked to them in a costless manner to complete a full $\hat{\omega}$-tree from the above set of links with a total cost of $r$. This completes the proof of the Lemma.

*Proof of Lemma 13.*   Let $\tilde{\omega} = (\tilde{x}, \tilde{x}, ..., \tilde{x})$, $\tilde{x} \neq x^w$, be some limit (monomorphic) state of the mutation-free dynamics. Every $\tilde{\omega}$-tree $\tilde{Y}$ must incur a cost $c(\tilde{Y}) \geq r$ since at least one mutation is needed to escape *every* one of the $r$

limit sets $\{\{\omega_x\} : x \neq \tilde{x}\}$. But, in fact, the cost of escaping the limit set $\{x^w\}$ must be no smaller than two. This follows from the derivations carried out in Subsection 2.7.2, which show that in a context where all but one firm produce output $x^w$, this single firm obtains a profit lower than the rest. Thus, at least two mutations are necessary to escape the limit set $\{x^w\}$, leading to $r + 1$ as the lower bound for $c(\tilde{Y})$. This completes the proof of the Lemma (and the Proposition). ■

As mentioned, the previous result illustrates the potential of evolutionary analysis for the study of classical economic problems. Theorem 21, in particular, provides a new evolutionary foundation for Walrasian behaviour which is very different from other traditional approaches found in the literature. These approaches – both those based on Co-operative and those based on Non-co-operative Game Theory[33] – have aimed at characterizing those circumstances in which, in a general sense, there is no "monopoly power" (i.e. no agent has the possibility of affecting his terms of trade). In essence, this has been seen to rely crucially on the existence of a large enough population, a consideration which is fully absent from the present analysis. (The conclusion obtained in Proposition 14 is independent of the number of firms in the market.)

---

[33] See e.g. Hildenbrand (1974) or Mas-Colell (1980) for a discussion of the co-operative and non-co-operative approaches respectively.

# 6
# Evolution, Expectations, and Drift

## 6.1 Introduction

The approach undertaken in the previous chapter explored a stochastic evolutionary framework which is a direct analogue of biological (Darwinian) processes. In particular, the postulated selection dynamics involved the requirement that any strategy which performs better (on average) should tend to see its frequency increase at the expense of other worse-performing strategies. Some simple, but appealing, assumptions on the underlying behaviour of agents would induce such a formulation. For example, one could simply assume (as in Section 5.7) that players tend to imitate those strategies whose average pay-off is highest.

There are, however, many economic applications where it is natural to allow for agents who rely on richer, more sophisticated, considerations in shaping their behaviour. In consonance with much of traditional economic analysis, a natural option in this respect is to consider agents who can form some (certainly *not* "rational") expectations on the future course of the system, then reacting optimally to them. Recent evolutionary literature largely focuses on what appears to be the first, most straightforward possibility of addressing this issue. It postulates that agents simply look at the *immediate* past and use it as a one-point predictor of what will happen next. Players, in other words, are assumed to form what is customarily labelled *static expectations*.

This approach encounters some difficulties if the immediate past does not necessarily provide a sufficient basis for the formation of a *complete* range of expectations. This will be the case, for example, in most extensive-form games where some of the information sets of the game may not have been reached in the previous period. Since many interesting economic applications will display this feature (recall, for example, the two-stage cheap talk games discussed in Section 2.8), an extension of the static-expectations approach will be needed to deal with this problem.

As a further extension, one would also want to consider the possibility that agents, aware that they are in an ever-changing context, may want to form dynamic (i.e. not necessarily static) expectations on the evolution of the process. Of course, in shaping such expectations, history should still continue to play an important role. Only if some requirement of "historical consistency" is imposed on the expectation-formation rule may one hope that the model will exhibit any interesting long-run regularities. However, in contrast with the strong rigidity induced by static expectations, "dynamic" expectations introduce new degrees of freedom whose implications are worth exploring in some detail.

The key new phenomenon which will be seen to play a crucial role when agents rely on dynamic expectations is some notion of (*expectational*) *drift*. In fact, a similar idea happens to underlie some of our discussion of extensive-form games also, *even* when the expectation-formation rule is static. The role of drift in these models is very analogous to that often attributed to genetic drift in biological contexts. As stressed by many biologists (see Kimura (1983)), drift enhances substantially the adaptability of the system by allowing for the build-up of *initially* neutral, but *later on* decisive, "material" for fruitful selection.

Our approach in this chapter will be as follows. First, in the next section, a general framework is proposed that encompasses all the different specific scenarios which are subsequently studied. Then, specializing this general framework in different directions (different updating rules, alternative types of games, some examples) a diverse set of issues is formally addressed: equilibrium selection and forward induction, bargaining, equilibrium volatility, etc. The chapter closes with a certain recapitulation which compares the different behavioural paradigms considered and proposes some stylized discussion on the "evolution of smartness", i.e. the issue of how (or whether) evolutionary processes may select among alternative degrees of players' sophistication.

## 6.2   General Theoretical Framework

In contrast with Chapter 5, the context here is assumed to involve two different populations (or types), 1 and 2, each assumed of equal size $n$.[1] In every period $t = 1, 2, ...$, members of both populations are assumed randomly matched in pairs (one from each population, some given number of independent rounds)[2] in order to play a certain bilateral game $G$ in extensive form with perfect recall.[3]

---

[1] The reason for contemplating two different populations here is to accommodate a wider number of applications which require this feature (see below). In cases where two populations are not indispensable (e.g. in Subsections 6.4.1 or 6.6.3), the results obtained carry over to the single-population scenario in an obvious fashion.

[2] The particular number of rounds played every period (possibly "infinite", which would eliminate all matching uncertainty) is inconsequential for the present approach.

[3] See e.g. Fudenberg and Tirole (1991: 77–82) for the standard formal description of a game in extensive form.

The (finite) strategy set of each population is respectively denoted by $S^i$, pay-off functions given by $\pi^i : S^1 \times S^2 \to \mathfrak{R}$.

Let $U^i$, $i = 1, 2$, denote the collection of information sets of each player type in game $G$, and let $A(u)$ stand for the set of actions available at each of its information sets $u \in U^i$. Every player of type $j = 3 - i$ is postulated to have a pattern of expectations (a collection of probability densities) over each of the sets $A(u)$, $u \in U^i$. These expectations specify the subjective probabilities which a player of type $j$ attributes to his opponent playing each of the actions available in the latter's information sets.

Expectations are taken to depend on history in the following two different ways.

(a) They are formulated as a *pattern* of expectations *conditional* on every possible past history. It is precisely such conditioning on history which will allow for genuinely dynamic expectations.

(b) Of course, expectations are also assumed *updated* on the basis of past history. This is reflected by the alternative postulates of "historical consistency" contemplated below.

Even though the approach undertaken here may be extended to the consideration of any finite strings of past history, nothing essential is lost for our purposes if, for the sake of simplicity, we restrict ourselves to two-period histories, i.e. histories which involve just the current and immediately preceding periods. Thus, simplifying the theoretical framework in this fashion, a *history* up to (and including) $t$ just requires the specification of which actions have been taken by all players in periods $t$ and $t - 1$. For each population $i = 1, 2$, denote

$$A^i \equiv \bigcup_{u \in U^i} A(u),$$

and

$$X^i \equiv \left\{ \left[ x^i(a) \right]_{a \in A^i} : x^i(a) \in \mathcal{N} \cup \{0\}, \, x^i(a) \leq n \right\}.$$

Then, the set $H$ of all possible (two-period) histories may be identified with an appropriate subset[4] of $\left[ X^1 \times X^2 \right] \times \left[ X^1 \times X^2 \right]$. For any given $t$, a typical element of it, $h_t = \left[ (x_t^1, x_t^2), (x_{t-1}^1, x_{t-1}^2) \right]$, includes the vectors $x_t^i$ and $x_{t-1}^i$ specifying, for each population $i = 1, 2$, how many individuals have played each action $a \in A^i$ in periods $t$ and $t - 1$, respectively.

For each population $i = 1, 2$, a *pattern of expectations* is defined as a collection

$$e^i = \{ e^i(\cdot \mid h) : h \in H \}$$

[4] If the game $G$ is simultaneous, then the set of all possible histories can be *exactly* identified with $(\hat{X}^1 \times \hat{X}^2) \times (\hat{X}^1 \times \hat{X}^2)$, where $\hat{X}^i = \{ \left[ x^i(a) \right]_{a \in A^i} \in X^i : \sum_{a \in A^i} x^i(a) = n \}$. In general, however, the restrictions on $X^i$ characterizing admissible histories must be tailored to the specific details of the extensive-form game under consideration.

which, contingent on *every* possible prior history $h \in H$, specifies the subjective probability $e^i(a \mid h)$ that a type-$i$ player currently attributes to his opponent choosing (or planning to choose) action $a \in A(u)$ if the corresponding information set $u \in U^j$ is reached. Note of course that, for any given $u \in U^j$, we must have

$$\sum_{a \in A(u)} e^i(a \mid h) = 1.$$

To remain within a finite set-up, it will be assumed that the values of any $e^i(\cdot \mid h)$ must belong to a certain (arbitrarily fine) grid, which always includes (at least) every point of the form $r/q$ where $r = 0, 1, 2, ..., q$ and $q = 1, 2, ..., n$. (That is, it includes all conceivable "fractions" associated with any subset of players in each population.)

Let $E^i$, $i = 1, 2$, denote the set of admissible patterns of expectations for population $i$ and denote by $\Psi^i \equiv S^i \times E^i$ the corresponding set of possible *individual characteristics* (i.e. pairs consisting of a strategy and an expectation pattern). For any given $t$, the current state of the process is identified with the tuple $\omega_t = \left[ h_t, \left( z_t^1(\psi^1) \right)_{\psi^1 \in \Psi^1}, \left( z_t^2(\psi^2) \right)_{\psi^2 \in \Psi^2} \right]$ which specifies the current history $h_t$ and the number of individuals $z_t^i(\psi^i)$ displaying every possible characteristic $\psi^i$ in each population $i = 1, 2$. The set of all such possible states will be denoted by $\Omega$.

The dynamics of the process may be decomposed in the following two sequential components: (i) expectation updating; (ii) strategy adjustment.

(i) *Expectation updating*

For each population $i = 1, 2$, expectation patterns are assumed updated on the basis of some (time-invariant) updating rule of the form

$$\hat{\mathbf{e}}^i = F^i(\mathbf{e}^i, h_{t-1}) \tag{6.1}$$

which specifies the expectation pattern held at $t$ by any given agent of population $i$ whose preceding pattern was $\mathbf{e}^i$, given that prior history is $h_{t-1}$. The specific requirements demanded from such $F^i(\cdot)$ will depend on the particular scenario under consideration. For example, if the expectations are assumed static and the game is simultaneous (Section 6.4 below), the mapping $F^i(\cdot)$ becomes a trivial projection. If, instead, the game and/or expectations are assumed to be of a dynamic nature (Sections 6.5 and 6.6), the updating rule will generally be a much more complicated object. Each of these possibilities is formulated and discussed in detail when introduced below.

(ii) *Strategy adjustment*

Every period $t$, each individual is assumed to enjoy a common and independent probability $p \in (0, 1)$ of being able to revise his strategy. In that event, and given any pattern of expectations $\mathbf{e}^i$ which the agent happens to hold at $t$, he chooses a new strategy which is a best response to the beliefs $e^i(\cdot \mid h_{t-1})$ induced by the prior history $h_{t-1}$. If, abusing previous notation, we simply denote by $\pi^i(s^i; \mathbf{e}^i, h_{t-1})$ the expected pay-off for strategy $s^i$ induced by the pattern of expectations $\mathbf{e}^i$ and history $h_{t-1}$, the set of such best responses at $t$ is given by

$$B^i(\mathbf{e}^i, h_{t-1}) = \{\hat{s}^i \in S^i : \pi^i(\hat{s}^i; \mathbf{e}^i, h_{t-1}) \geq \pi^i(s^i; \mathbf{e}^i, h_{t-1}), \forall s^i \in S^i\}.$$

If this set is not a singleton, all of its elements are assumed chosen, say, with equal probability, independently across players.

Strategy adjustment and expectation updating embody the selection component of the evolutionary dynamics. As before, it will be postulated that, before play is conducted in every period,[5] each individual is subject to a common independent probability $\epsilon > 0$ of mutation. In this event, his new characteristic is selected from the corresponding set $\Psi^i$ according to some given (time-invariant) probability distribution with full support.

Mutation, as before, will have the effect of making the process ergodic. Its unique invariant distribution – an element of $\Delta(\Omega)$ – will be denoted by $\mu_\epsilon$, the focus being again on its limit when $\epsilon$ becomes arbitrarily small.

## 6.3 Static Expectations

Young (1993$a$) was the first to introduce a process of expectation formation into stochastic evolutionary models in an explicit manner. His framework is somewhat different from the one proposed above. Specifically, he considers a context where, every period, only two individuals are randomly drawn to play a certain bilateral game. In deciding what to do, each of these two players looks back at history in order to shape his expectations on the opponent's strategy. Both of them are assumed to obtain a finite (non-exhaustive) sample of past (finite)[6] history. Once this sample is obtained, they have their subjective probabilities simply match the sample frequencies, then playing a best response to these expectations.

In contrast with the theoretical framework presented above, Young's sampling-based formulation introduces a mechanism for expectation formation that is of a stochastic nature. But except for this feature (which could be made consistent with our general framework by postulating a random updating function in (6.1)), Young's approach essentially reflects a static-like updating rule. To analyse in

---

[5] This formulation of mutation is proposed in order to keep the formal parallelism with what was postulated in the previous chapter. However, many other specifications are possible e.g. a natural one in the present context would be to restrict mutation to operate exclusively on the expectation component of individuals' characteristics.

[6] As before, history is truncated in order to preserve the finite and Markovian nature of the process, any past realization of the process beyond a certain point taken to be either unavailable or ignored.

a simple fashion the implications of such a static-expectations approach within our proposed framework, we postulate:

**Static Expectations (SE)**   Given any history $h = \left[(x_0^1, x_0^2), (x_1^1, x_1^2)\right] \in H$ and some former pattern of expectations $\mathbf{e}^i \in E^i$ ($i = 1, 2$) the new (updated) expectation pattern

$$\hat{\mathbf{e}}^i = F^i(\mathbf{e}^i, h)$$

satisfies the following condition:   $\forall a \in A(u),\ u \in U^j,\ j = 3 - i,$

$$
\begin{aligned}
\hat{e}^i(a \mid h) &= \frac{x_0^j(a)}{\sum_{a' \in A(u)} x_0^j(a')} && \text{if } \sum_{a' \in A(u)} x_0^j(a') > 0\ ; \\
\hat{e}^i(a \mid h) &= e(a \mid h) && \text{otherwise.}
\end{aligned}
$$

Moreover, $\hat{e}^i(\cdot \mid \tilde{h}) = e^i(\cdot \mid \tilde{h})$, for all $\tilde{h} \neq h$.

Thus, verbally, the former postulate specifies that agents update their relevant expectations (i.e. those contingent on prevailing history) by simply "matching" the action frequencies just *observed* at every information set. If no instance of play is observed at some information set, the previous associated beliefs remain unchanged. Notice that this formulation implicitly assumes that agents carry out their considerations independently at each information set. Thus, in particular, they are unable to detect any correlation of behaviour across them. If such correlation could be observed, the above formulation should be modified, this having no essential effect on our future analysis.

## 6.4    Simultaneous Contexts

### 6.4.1    Co-ordination games

To explore the implications of SE, we start by focusing on the simple co-ordination case studied in Section 5.4. Suppose that every pair of matched individuals plays a symmetric[7] ($2 \times 2$)-game as described by Table 15 with

$$a > c,\ b > d.$$

In line with the conclusions obtained by Young (1993a), the next result shows that Postulate SE above (i.e. static expectations) yields the long-run selection of the risk-dominant equilibrium.

**Theorem 22**  *Assume SE, $c + b > a + d$, and $n$ even. Let $h = \big[(x_k^1(s_1), x_k^1(s_2)),$ $(x_k^2(s_1), x_k^2(s_2))\big]_{k=0,1}$ represent a typical history of the process. Then, any*

---

[7] Even though the present two-population context would allow the study of asymmetric games, this is postponed to Subsection 6.4.2 below. Here, the focus is on the same symmetric context studied in Ch. 5 in order to facilitate the comparison with previous analysis.

given $\tilde{\omega} = \left[ \tilde{h}, \left( \tilde{z}^i(\psi^i) \right)_{\psi^i, i} \right] \in \Omega$ *displays* $\lim_{\epsilon \to 0} \mu_\epsilon(\tilde{\omega}) > 0$ *if, and only if,*
$\tilde{h} = [(0, n), (0, n)]^2$ *and*

$$\tilde{z}^i(s^i, \mathbf{e}^i) > 0 \; (i = 1, 2) \Rightarrow \left[ s^i = s_2 \; \wedge \; e^i(s_2 \mid \tilde{h}) = 1 \right].$$

**Proof.** Despite their different frameworks, the proof here is quite similar to that of Theorem 20. In order to emphasize their parallelisms, we use the same notation throughout, even though the objects to which it refers are obviously different. No confusion should arise.

Let $T_\epsilon$ be the transition matrix of the evolutionary process induced by some given $\epsilon > 0$. Distance and cost functions on $\Omega \times \Omega$ are defined as follows:[8]

$$\begin{aligned} d(\omega, \hat{\omega}) \quad &= \tfrac{1}{2} \sum_{\substack{\psi^i \in \Psi^i \\ i=1,2}} \left| z^i(\psi^i) - \hat{z}^i(\psi^i) \right| \quad \text{if } T_\epsilon(\omega, \hat{\omega}) > 0 \\ &= \infty \qquad\qquad\qquad\qquad\qquad \text{otherwise}, \end{aligned} \tag{6.2}$$

where note that, unlike the case of the scenario studied in Chapter 5, one may well now have $T_\epsilon(\omega, \hat{\omega}) = 0$ if the histories included in $\omega$ and $\hat{\omega}$ are not consistent, i.e. if the first component of $\omega$ (i.e. its "current" outcome) is not identical to the second component of $\hat{\omega}$ (i.e. its "former" outcome). Based on the function $d(\cdot, \cdot)$, a cost function for each possible transition is defined as follows:

$$c(\omega, \hat{\omega}) = \min_{\omega' \in \Omega} \{ d(\omega', \hat{\omega}) : T_0(\omega, \omega') > 0 \}, \tag{6.3}$$

where, as before, $T_0(\cdot)$ stands for the transition-probability matrix of the mutation-free dynamics.

Let $h_1 \equiv [(n, 0), (n, 0)]^2$ and $h_2 \equiv [(0, n), (0, n)]^2$, i.e. the constant histories induced by the two populations *all* playing either one strategy or the other for two consecutive periods. For each strategy, $s_1$ and $s_2$, and each population $i = 1, 2$, define

$$\begin{aligned} \Psi_1^i \quad &\equiv \{ \psi^i = (s^i, \mathbf{e}^i) : s^i = s_1 \; \wedge \; e^i(s_1 \mid h_1) = 1 \} \\ \Psi_2^i \quad &\equiv \{ \psi^i = (s^i, \mathbf{e}^i) : s^i = s_2 \; \wedge \; e^i(s_2 \mid h_2) = 1 \} \end{aligned} \tag{6.4}$$

and then consider, for each $q = 1, 2$, the following two subsets of the state space:

$$\Upsilon_q \equiv \{ \omega = \left[ h, \left( z^i(\psi^i) \right)_{\psi^i, i} \right] \in \Omega : h = h_q \; \wedge \sum_{\psi^i \in \Psi_q^i} z^i(\psi^i) = n, \; \forall i = 1, 2 \}. \tag{6.5}$$

Only states in either $\Upsilon_1$ or $\Upsilon_2$ can belong to absorbing sets of the mutation-free dynamics. To verify this, it is enough to confirm that, from any state $\omega_t$ prevailing at some $t$, there is positive probability (bounded above zero)

---

[8] The interpretation here bears a close analogy to that given before, i.e., provided there is a feasible transition from $\omega$ to $\hat{\omega}$, $d(\omega, \hat{\omega})$ indicates the number of agents whose characteristics differ between the former and the latter states. Note that the factor $\tfrac{1}{2}$ simply eliminates double counting.

that the mutation-free process enters one of these two sets in some prespecified number of periods and remains there. Suppose, for example, that each of the two populations fully adjusts in turn (say, first the whole of population 1 at $t+1$, then everyone of population 2 at $t+2$). This pattern has positive probability, bounded above zero, because $p \in (0, 1)$. Since every individual shares the same relevant expectations at the time of adjustment, it may be assumed that (at least with some given positive probability) every one of them in each population adjusts towards the same strategy. This, at $t+2$, leads to a state where every agent in both populations chooses the same strategy. From this state, that prevailing in the ensuing period $t+3$ must belong to either $\Upsilon_1$ or $\Upsilon_2$. But then, since every state in either $\Upsilon_1$ or $\Upsilon_2$ is stationary in the mutation-free dynamics, the process will trivially remain in each of these sets in the absence of mutation. This, incidentally, also confirms the validity of the statement reciprocal to the one initially aimed at. That is, every state in $\Upsilon_1 \cup \Upsilon_2$ belongs to an absorbing set (in fact, a singleton) of the mutation-free dynamics.

From the usual considerations, the proof of the theorem then follows from the two following Lemmas. Let $r = card(\Upsilon_1) + card(\Upsilon_2)$.

**Lemma 14** *There exists some $\hat{\omega} \in \Upsilon_2$ with an $\hat{\omega}$-tree $\hat{Y}$ such that $c(\hat{Y}) \leq r - 2 + \frac{n}{2}$.*

**Lemma 15** *For any $\hat{\omega} \in \Upsilon_1$, every $\hat{\omega}$-tree $\hat{Y}$ has $c(\hat{Y}) > r - 2 + \frac{n}{2}$.*

*Proof of Lemma 14:*   Let $\omega \in \Upsilon_1$. First, it is argued that there is a four-link path $y = \{(\omega, \omega'), (\omega', \omega''), (\omega'', \omega'''), (\omega''', \hat{\omega})\}$ leading from $\omega$ to some state $\hat{\omega} \in \Upsilon_2$ such that $c(y) = \frac{n}{2}$. To verify this claim, choose $\omega'$ to differ from state $\omega$ only in that $\frac{n}{2}$ individuals of one of the populations, say population 1, mutate towards strategy $s_2$, expectations then being adjusted according to the updating rule. From (6.3), $c(\omega, \omega') = \frac{n}{2}$. Then, from state $\omega'$ the mutation-free dynamics attributes positive probability to the transition towards a state $\omega''$ where all individuals of population 1 stay with their former strategy – due to inertia, whose (marginal) probability is $(1 - p)^n$ – but *all* individuals of population 2 have the opportunity of adjusting their strategy. In that case, the assumption that $c + b > a + d$ implies that all of them will shift to strategy $s_2$. This transition has therefore a cost $c(\omega', \omega'') = 0$. Then, since the transition towards the state $\omega'''$ where the remaining $\frac{n}{2}$ individuals of population 1 switch to $s_2$ also has positive probability according to the mutation-free dynamics, it also follows that $c(\omega'', \omega''') = 0$. Finally, if $\hat{\omega}$ simply denotes the state following $\omega'''$ when no individual mutates (and therefore $c(\omega'', \hat{\omega}) = 0$), $\hat{\omega} \in \Upsilon_2$ as desired.

To complete the proof of the Lemma, one just needs to observe that the path $y$ just constructed can be completed to form a full $\hat{\omega}$-tree at an additional cost of $r - 2$. First, notice that *all* states in $\Upsilon_1 \backslash \{\omega\}$ can be directly or indirectly linked to $\omega$ with a cost equal to $card(\Upsilon_1) - 1$ through a sub-tree $\bar{Y}$ (restricted to $\Upsilon_1$) which involves only one-step mutations. This is due to the fact that all states in $\Upsilon_1$ only differ in expectations contingent on "unobserved" histories.

Analogously, all states in $\Upsilon_2\backslash\{\hat{\omega}\}$ can be directly or indirectly linked to $\hat{\omega}$ with a cost equal to $card(\Upsilon_2) - 1$ through a sub-tree $\check{Y}$ on $\Upsilon_2$ with only one-step mutations. Since all the remaining states in $\Omega\backslash(\Upsilon_1 \cup \Upsilon_2)$ are just transitory in the mutation-free dynamics, it is clear that $y \cup \bar{Y} \cup \check{Y}$ can be further completed to form a full $\hat{\omega}$-tree at no extra cost, i.e. with a *total* cost of $c(\bar{Y}) + c(\check{Y}) + c(y) = r - 2 + \frac{n}{2}$. This completes the proof of Lemma 14.

*Proof of Lemma 15:* Let $\hat{\omega} \in \Upsilon_1$ and suppose, for the sake of contradiction, that there exists some $\hat{\omega}$-tree $\hat{Y}$ whose cost $c(\hat{Y}) \leq r - 2 + \frac{n}{2}$. First note that since every state in $\Upsilon_1 \cup \Upsilon_2$ is a stationary state of the mutation-free dynamics, at least one mutation is required to exit it. This implies that every $\hat{\omega}$-tree must incur a cost no smaller than $r - 1$ (i.e. at least a cost of 1 associated with each of the arrows originating in the states belonging to $(\Upsilon_1 \cup \Upsilon_2)\backslash\{\hat{\omega}\}$). On the other hand, observe as well that any $\hat{\omega}$-tree must involve a path $y$ linking some state in $\Upsilon_2$ to some other state in $\Upsilon_1$. Since it is immediately seen that any such path must have a cost $c(y) \geq \frac{n}{2} + 1$, a contradiction results. This completes the proof of the Lemma and the Theorem. ∎

In line with Theorem 20, the previous result establishes a role for risk-dominance as a criterion for equilibrium selection in simple $(2 \times 2)$-games when players optimize relative to static expectations. As will be discussed below (Subsection 6.5.2), a certain "dynamic" enrichment of the former scenario (one which allows for a prior outside option) leads back to efficiency as the selection criterion in simple co-ordination games – recall Theorem 19.

One particular aspect of the previous result is worth some attention, since it will reappear below with much more substantial implications. Even though Theorem 22 fully narrows down long-run behaviour, it leads nevertheless to a large degree of "expectation indeterminacy". Specifically, it only imposes constraints on those beliefs which are contingent on established long-run history. For any other histories, no discipline is imposed on the pattern of expectations since, naturally, the updating process can only be responsive to those histories which are being observed.

In principle, therefore, one may well have a substantial range of expectational heterogeneity due to drift (i.e. mutations unopposed by selection forces). In the present context, however, such heterogeneity is inconsequential because static updating brings in *immediate* convergence of the expectation patterns where it matters, namely on the unique information set of each population (always visited along any path of play), and contingent on the only relevant history (that recently observed). As discussed below, the implications of this potential heterogeneity will turn out to be important when either the game is dynamic (i.e. it involves sequential decisions, as in Section 6.5) or expectations are dynamic (i.e. do not necessarily predict a stationary evolution of the process, as in Section 6.6).

### 6.4.2    A simple model of bargaining

The analysis conducted so far (both in this chapter and the previous one) has focused solely on symmetric games. However, by allowing now for two distinct populations, we have the potential of analysing interesting asymmetric interactions between them. To illustrate these new possibilities, we discuss here a simple bargaining scenario which is largely inspired in the much more elaborate model studied by Young (1993$b$).

Suppose that when two individuals of population 1 and 2 meet, they confront a restricted Nash demand game with just two proposals: a high one $M$ and a low one $N$. For simplicity, let us normalize the total surplus to be divided and make it equal to 2 units, identifying the high demand $M$ with some $x \in (1, \frac{3}{2})$ and the low demand $N$ with $2 - x < 1$. If the (simultaneous) proposals submitted are consistent (i.e. add up to less than 2), then each individual gets what he proposed plus half of any remaining surplus. Otherwise, if the proposals are inconsistent (their sum exceeds 2), no individual gets any surplus at all.

The amount of the surplus obtained represents the argument of each individual's utility function. Let $v^i : [0, 2] \to \Re$ stand for the common utility function of the individuals of population $i = 1, 2$. It is assumed to be of the following form:

$$v^i(r) = r^{\alpha^i}, \quad 0 < \alpha^i \leq 1.$$

Given any pair of utility functions, $v^1(\cdot)$ and $v^2(\cdot)$, the bilateral game played by each pair of individuals can be represented as follows:

| 2 \ 1 | $M$ | $N$ |
|---|---|---|
| $M$ | $0, 0$ | $x^{\alpha^1}, (2 - x)^{\alpha^2}$ |
| $N$ | $(2 - x)^{\alpha^1}, x^{\alpha^2}$ | $1, 1$ |

Table 16

This game has only two pure-strategy equilibria, $(M, N)$ and $(N, M)$, which are asymmetric. Moreover, the game itself is asymmetric if $\alpha^1 \neq \alpha^2$. Conceiving of each population's $\alpha^i$ as a measure of their respective risk aversion, one would intuitively expect that the population with the higher such parameter (i.e. the population which is less "risk-averse") would end up capturing the larger share of the surplus in the long run. Or if, say, $\alpha^1 > \alpha^2$, the long-run outcome should be concentrated in the equilibrium $(M, N)$ which favours population 1. Modelling the evolutionary process as above, this conjecture is indeed confirmed (provided the populations are large enough) by the following result.

**Proposition 15** *Assume SE and $\alpha^1 > \alpha^2$, and let $h = [(x_k^1(M), x_k^1(N)),$ $(x_k^2(M), x_k^2(N))]_{k=0,1}$ be a typical history of the process. There exists some $\tilde{n} \in \mathcal{N}$ such that if $n \geq \tilde{n}$, any given $\tilde{\omega} = \left[\tilde{h}, (\tilde{z}^i(\psi^i))_{\psi^i,i}\right] \in \Omega$ displays $\lim_{\epsilon \to 0} \mu_\epsilon(\tilde{\omega}) > 0$ if, and only if, $\tilde{h} = [(n,0),(0,n)]^2$ and*

$$\tilde{z}^1(s^1, \mathbf{e}^1) > 0 \Rightarrow \left[s^1 = M \wedge e^1(N \mid \tilde{h}) = 1\right],$$
$$\tilde{z}^2(s^2, \mathbf{e}^2) > 0 \Rightarrow \left[s^2 = N \wedge e^2(M \mid \tilde{h}) = 1\right].$$

**Proof.** Define by $\gamma^i \in (0,1)$ the fraction of population $i = 1,2$ which has to play strategy $M$ in order for individuals in population $j = 3 - i$ to be indifferent between either strategy, $M$ or $N$ – that is, the probability-weights defining the unique mixed-strategy equilibrium of the game. They are easily computed to be:

$$\gamma^i = \frac{x^{\alpha^j} - 1}{x^{\alpha^j} - 1 + (2-x)^{\alpha^j}} \qquad i, j = 1, 2, \ j = 3 - i. \tag{6.6}$$

By differentiating the above expression with respect to $\alpha^j$, one may verify that it is increasing in $\alpha^j$, ranging from $\gamma^i = 0$ if $\alpha^j = 0$ to $\gamma^i = x - 1$ if $\alpha^j = 1$. Thus, in particular, since it is assumed that $\alpha^1 > \alpha^2$, it follows that $\gamma^1 < \gamma^2$.

Define, for each population $i = 1, 2,$[9]

$$\Psi_M^1 \equiv \{\psi^1 = (s^1, \mathbf{e}^1) : s^1 = M \wedge e^1(N \mid [(n,0),(0,n)]^2) = 1\}$$
$$\Psi_M^2 \equiv \{\psi^2 = (s^2, \mathbf{e}^2) : s^2 = M \wedge e^2(N \mid [(0,n),(n,0)]^2) = 1\},$$

which specify the set of characteristics in which the respective population plays strategy $M$, expecting the other population to play $N$ contingent on a history which fully matches this behaviour. The sets $\Psi_N^1$ and $\Psi_N^2$ may be defined analogously. Based on these sets (which are the counterparts of those defined in (6.4) for the present context), the analogues of (6.5) are defined as follows:

$$\Upsilon_{MN} \equiv \left\{ \begin{array}{l} \omega = \left[h, (z^i(\psi^i))_{\psi^i,i}\right] \in \Omega : h = [(n,0),(0,n)]^2 \ \wedge \\ \sum_{\psi^1 \in \Psi_M^1} z^1(\psi^1) = \sum_{\psi^2 \in \Psi_N^2} z^2(\psi^2) = n \end{array} \right\}$$

$$\Upsilon_{NM} \equiv \left\{ \begin{array}{l} \omega = \left[h, (z^i(\psi^i))_{\psi^i,i}\right] \in \Omega : h = [(0,n),(n,0)]^2 \ \wedge \\ \sum_{\psi^1 \in \Psi_N^1} z^1(\psi^1) = \sum_{\psi^2 \in \Psi_M^2} z^2(\psi^2) = n \end{array} \right\}.$$

The previous sets represent the two *state components* where the populations indefinitely play, and expect to play, one of the two pure-strategy equilibria. Along the lines of the proof of Theorem 22, it can easily be verified that a certain state belongs to some absorbing set of the mutation-free dynamics if, and only if, it is an element of one of these sets. Thus, relying on familiar considerations, the

---

[9] In terms of previous notation, proposal $M$ is interpreted here as strategy $s_1$ and proposal $N$ as strategy $s_2$.

proof of the present result follows from the following two Lemmas. (Previous notation, when obviously reinterpretable, remains unchanged throughout.)

**Lemma 16** *For any $\hat{\omega} \in \Upsilon_{NM}$, there is a path $y$ linking $\hat{\omega}$ to some $\breve{\omega} \in \Upsilon_{MN}$ with $c(y) \leq \left[\gamma^1 n\right]^+$, where $[x]^+$ stands for the smallest integer no smaller than $x$.*

*Proof:* Consider any $\hat{\omega} \in \Upsilon_{NM}$, and let $\omega'$ be the state derived from $\hat{\omega}$ by $\left[\gamma^1 n\right]^+$ mutations in population 1 towards strategy $M$ and the corresponding adjustment on expectations. This transition has a cost $c(\hat{\omega}, \omega') = \left[\gamma^1 n\right]^+$. Now let $\omega''$ be the state reached from $\omega'$ when *all* individuals of population 2 switch to strategy $N$ but *no* member of population 1 receives a strategy-adjustment opportunity. From the definition of $\gamma^1$ in (6.6), this transition can be chosen to have a cost $c(\omega', \omega'') = 0$. Then, consider the transition to the state $\omega'''$ which results when all members of population 1 are given the opportunity to adjust their strategies. Again, $c(\omega'', \omega''') = 0$, and in this state every agent of population 1 chooses $M$ and everyone in population 2 chooses $N$. Without any further mutation, the ensuing state $\breve{\omega} \in \Upsilon_{MN}$, which completes the proof of the Lemma.

**Lemma 17** *For any $\hat{\omega} \in \Upsilon_{MN}$, every path $y$ linking $\hat{\omega}$ to some $\breve{\omega} \in \Upsilon_{NM}$ has a cost $c(y) \geq \left[\gamma^2 n\right]^+$.*

*Proof:* First, note that since

$$\gamma^1 < \gamma^2 \leq x - 1 < \tfrac{1}{2}$$

we must have

$$\gamma^2 < 1 - \gamma^1.$$

This indicates that the "easiest" transition from a state in $\Upsilon_{MN}$ to a state in $\Upsilon_{NM}$ must involve members of population 2 first mutating towards strategy $M$. Any fewer than $\left[\gamma^2 n\right]^+$ mutations from any state $\hat{\omega} \in \Upsilon_{MN}$ still has the process remain in the basin of attraction of the set $\Upsilon_{MN}$. (That is, if no further mutation occurs, the selection dynamics will lead the process back to some state in $\Upsilon_{MN}$ with probability one.) Therefore, in order to complete a transition from $\hat{\omega}$ to some $\breve{\omega} \in \Upsilon_{NM}$ the corresponding path $y$ must involve at least $\left[\gamma^2 n\right]^+$ mutations. This integer is then a lower bound on $c(y)$, which is the desired claim.

To complete the proof of the Proposition, simply note that the inequality $\gamma^1 < \gamma^2$ implies $\left[\gamma^1 n\right]^+ < \left[\gamma^2 n\right]^+$ for large enough $n$.  ∎

Proposition 15 reflects in, a very schematic way, considerations very similar to those established by Young (1993*b*) in a model where the range of proposals consists of an arbitrarily fine grid. Applying to this richer framework the sampling approach to expectation formation which was outlined in Section 6.3, Young shows that the unique long-run outcome of the process approximates the Nash bargaining solution (Nash (1950)) as the grid of proposals becomes arbitrarily fine.[10] The conclusion of Proposition 15 is a very simplified variation on Young's result. As in the Nash bargaining solution, the outcome selected by the evolutionary process provides the largest share of the surplus to the less risk-averse party.

## 6.5 Multi-Stage Contexts

### 6.5.1 Introduction

When the game being played is simultaneous (as in the previous section), a static rule for expectation formation imposes an immediate homogeneity on players' expectations at the opponents' unique information set. No *relevant* heterogeneity in expectations can arise within each population since any past history has been commonly observed. This, of course, is not any longer true when the game in question displays a genuine dynamic dimension and, therefore, some paths of past play may have failed to visit certain information sets. In this case, substantial expectation heterogeneity may set in at these information sets, leading in turn to some significant implications on the evolution of the process.

To illustrate these matters in a simple way, we turn again to $(2 \times 2)$-co-ordination games. This set-up is now embedded into a dynamic scenario by providing one of the parties with an outside option to be exercised (or not) prior to the co-ordination game. Despite the fact that this outside option ends up *not* being used in the long run, its mere availability turns out to destabilize the inefficient equilibrium. In fact, the considerations involved in this respect happen to be quite reminiscent of the familiar ideas of forward induction often used in classical Game Theory (recall Subsection 4.4.2). Here, however, the implicit (and iterative) reasoning process typically conceived to underlie it does not take place in the players' minds but is the "patient" work of evolutionary drift.

---

[10] Strictly speaking, Young (1993*b*) shows that the outcome will approximate the usual (symmetric) Nash Bargaining Solution only if, except for their pay-off functions, the two populations are fully symmetric. He proves, for example, that if one population samples a larger fraction of past history than the other one, the long-run outcome will be biased in the former's favour. In this case, the outcome coincides with the asymmetric (or weighted) Nash bargaining solution, where each population's weights correspond to their respective sampling fractions.

In our context, one can also incorporate some natural asymmetries not derived from differences in the utility functions, e.g. it is easy to see that, if the two populations have a different cardinality, the one with fewer members will be favoured by the long-run outcome (in particular, if both populations have the *same* pay-off functions, the one with fewer individuals will achieve the $x$-share of the surplus in the long run).

### 6.5.2   *Forward induction and efficient co-ordination*

Assume that two populations, 1 and 2, are matched to play a co-ordination game as described in Subsection 6.4.1. Now, however, before entering the co-ordination game proper, let us assume that the player of population 1 may choose to avoid playing this game and instead ensure for himself a certain pay-off equal to $f$. If he does so, the type-2 player gets a pay-off of $g$, whose magnitude is fully irrelevant to what follows.

One possible extensive-form representation of the game is presented in Figure 10. In this extensive-form game, the type-1 player has two information sets: $u_1^1$ at the beginning of the game, $u_2^1$ after player 2's move. Thus, his (pure) strategy set $S^1$ can be identified with the Cartesian product $\{P, Q\} \times \{s_1, s_2\}$, where $P$ and $Q$ stand for "play" and "quit" respectively, the actions available in his first information set. On the other hand, the strategy set for a type-2 player merely coincides with that of the co-ordination game, $S^2 = \{s_1, s_2\}$, since his only information set is within the co-ordination subgame.

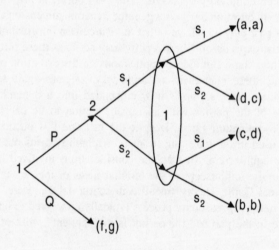

Figure 10: A coordination game with an outside option

To make the considerations involved interesting, suppose that

$$a > f > b. \tag{6.7}$$

That is, the "outside option" for player 1 guarantees him a pay-off which is precisely between the two different pay-offs attainable at each of the respective equilibria of the ensuing co-ordination (sub-)game. The next result establishes that, under static expectation-updating (Postulate SE above), every long-run state of the process has agents playing the co-ordination subgame *and* its efficient equilibrium. This conclusion requires that the population be large enough, but is independent of any other (in particular, risk-dominance) considerations.[11]

**Proposition 16 (Nöldeke and Samuelson (1993))** *Assume SE and (6.7), and let* $h = \left[(x_k^1(Q), x_k^1(P), x_k^1(s_1), x_k^1(s_2)), (x_k^2(s_1), x_k^2(s_2))\right]_{k=0,1}$ *be a typical history of the process. There exists some* $\tilde{n} \in N$ *such that if* $n \geq \tilde{n}$, *every* $\tilde{\omega} = \left[\tilde{h}, \left(\tilde{z}^i(\psi^i)\right)_{\psi^i,i}\right] \in \Omega$ *with* $\lim_{\epsilon \to 0} \mu_\epsilon(\tilde{\omega}) > 0$ *satisfies* $\tilde{h} = [(0, n, n, 0), (n, 0)]^2$ *and*

$$\tilde{z}^1(s^1, \mathbf{e}^1) > 0 \; \Rightarrow \; \left[s^1 = (P, s_1) \; \wedge \; e^1(s_1 \mid \tilde{h}) = 1\right],$$
$$\tilde{z}^2(s^2, \mathbf{e}^2) > 0 \; \Rightarrow \; \left[s^2 = s_1 \; \wedge \; e^2(P \mid \tilde{h}) = e^2(s_1 \mid \tilde{h}) = 1\right].$$

**Proof.** Denote by $h_Q \equiv [(n, 0, 0, 0), (0, 0)]^2$ and $h_P \equiv [(0, n, n, 0), (n, 0)]^2$ the two constant histories induced by each of the two alternative pure-strategy, Nash-equilibrium paths of play in the game. In the first one, $h_Q$, population 1 chooses to avoid the co-ordination subgame. In the second one, $h_P$, the whole of population 1 enters the co-ordination subgame to play its efficient equilibrium. Now, in analogy with (6.4), define:

$$\Psi_Q^1 \equiv \left\{ \begin{array}{c} \psi^1 = (s^1, \mathbf{e}^1) : \left[s^1 = (Q, s_q), q = 1, 2\right] \wedge \\ \left[\pi^1(s^1; \mathbf{e}^1, h_Q) > \pi^1(\hat{s}^1; \mathbf{e}^1, h_Q), \forall \hat{s}^1 \neq (Q, \cdot)\right] \end{array} \right\},$$
$$\Psi_P^1 \equiv \{\psi^1 = (s^1, \mathbf{e}^1) : s^1 = (P, s_1) \; \wedge \; e^1(s_1 \mid h_P) = 1\},$$
$$\Psi_P^2 \equiv \{\psi^2 = (s^2, \mathbf{e}^2) : s^2 = s_1 \; \wedge \; e^2(s_1 \mid h_P) = 1\}.$$

As for (6.5), one may now associate with the above subsets of individual characteristics the following two state components:

$$\Upsilon_Q = \{\omega = \left[h_Q, \left(z^i(\psi^i)\right)_{\psi^i,i}\right] \in \Omega : \sum_{\psi^1 \in \Psi_Q^1} z^1(\psi^1) = n\},$$
$$\Upsilon_P = \{\omega = \left[h_P, \left(z^i(\psi^i)\right)_{\psi^i,i}\right] \in \Omega : \sum_{\psi^1 \in \Psi_P^1} z^1(\psi^1) = \sum_{\psi^2 \in \Psi_P^2} z^2(\psi^2) = n\}.$$

The proof of the proposition follows from the following four Lemmata.

**Lemma 18** *A state* $\omega \in \Omega$ *belongs to an absorbing set of the mutation-free dynamics if, and only if,* $\omega \in \Upsilon_Q \cup \Upsilon_P$.

*Proof:* The argument is similar to that applied above. The sufficiency statement is obvious. For the necessity part, choose *any* arbitrary $\omega_t \in \Omega$

---

[11] The analysis here follows that of Nöldeke and Samuelson (1993) very closely. However, the framework is slightly different. For example, these authors postulate that when inertia operates on a particular agent, it "freezes" adjustment on *both* his strategy and his expectation.

prevailing at some $t$ and consider five consecutive periods $t+1, ..., t+5$ with the following alternative sequences of mutation-free events.

First, suppose that in the initial two periods $t+1$ and $t+2$, the whole of population 1 is given the opportunity to adjust. In implementing their adjustment, assume that every individual of population 1 who finds it *weakly* optimal to play the co-ordination subgame does enter it. Then, if even under these conditions neither of them decides to play the co-ordination subgame at $t+1$, the state $\omega_{t+2}$ belongs to $\Upsilon_Q$ and is stationary for the mutation-free dynamics.

Alternatively, suppose that some individual of population 1 does play $P$ at $t+1$. Then, in period $t+2$, since all individuals of population 1 share the same beliefs, suppose that *all* of them either play $Q$ or, if they *all* play $P$, then they choose the *same* ensuing action, $s_1$ or $s_2$.[12] In the former case, we may conclude that $\omega_{t+3} \in \Upsilon_Q$, as before. If, instead, the second possibility applies, suppose that players of population 1 are unable to adjust their strategy at $t+3$ but the whole of population 2 is able to do so. Then, the latter will all adjust to either $s_1$ or $s_2$, matching the single action chosen by population 1. In the first subcase (i.e. if they adjust towards $s_1$), the state $\omega_{t+4} \in \Upsilon_P$. In the second subcase (i.e. population 2 adjusts towards $s_2$ in $t+3$), every member of population 1 who could adjust his strategy at $t+4$ will switch to $Q$. Suppose indeed that all of them enjoy such adjustment possibility. Then, $\omega_{t+5} \in \Upsilon_Q$.

One of the different alternative paths considered has positive probability (bounded above zero). Thus, overall, it may be concluded that there is positive probability (bounded above zero) that, in at most five periods, the process reaches $\Upsilon_Q$ or $\Upsilon_P$. This obviously guarantees the desired conclusion.

**Lemma 19** *Let* $\omega, \omega' \in \Upsilon_r$ *for some given* $r = P, Q$. *Then,*

$$\min_{Y \in \mathcal{Y}_\omega} c(Y) = \min_{Y \in \mathcal{Y}_{\omega'}} c(Y). \tag{6.8}$$

*Proof:*   Let $\omega, \omega' \in \Upsilon_r$ be two different states such that $d(\omega, \omega') = 1$. From the definition of the sets $\Upsilon_r$, $r = P, Q$, this implies that the only difference between $\omega$ and $\omega'$ is that one (and only one) agent has expectation patterns in these states which differ *exclusively* for histories $h \neq h_r$.

Let $Y$ be some arbitrary $\omega$-tree. It is first claimed that there always exists some $\omega'$-tree $Y'$ such that $c(Y') \leq c(Y)$. The argument is constructive. Consider the arrow starting at $\omega'$ in tree $Y$, say $(\omega', \omega'')$. Delete this arrow and add the arrow $(\omega, \omega')$. These transformations produce an $\omega'$-tree whose cost cannot be larger than $c(Y)$. The reason is that the elimination of the arrow $(\omega', \omega'')$ in the tree $Y$ decreases the cost in at least one unit. On the other hand, the arrow $(\omega, \omega')$ has a cost $c(\omega, \omega') = 1$.

Since the argument is fully symmetric between $\omega$ and $\omega'$, one may in fact conclude that the equality (6.8) applies for any two such "neighbouring" states.

---

[12] Note that if $c > f$ ($> b$), players' beliefs may be such that it is optimal to enter the co-ordination subgame *and* play $s_2$.

But then, the conclusion of the lemma directly follows from the simple observation that every two states in a given $\Upsilon_r$ can be joined through a path whose links involve only neighbouring states (in the previous sense). This implies (6.8) for any two states in each $\Upsilon_r$, as desired.

**Lemma 20** *There exists a path $y$ linking some $\omega \in \Upsilon_Q$ to some other $\hat{\omega} \in \Upsilon_P$ with $c(y) = 1$.*

   *Proof:*  Consider the state $\omega = \left[ h_Q, \left(z^i(\psi^i)\right)_{\psi^i, i} \right] \in \Upsilon_Q$ where

$$z^2(s^2, e^2) > 0 \Rightarrow \forall h \in H, \ e^2(s_1 \mid h) = 1.$$

Consider now a state $\tilde{\omega} \in \Omega$ in which one of the individuals in population 1 has a pattern of expectations $e^1$ for which

$$\pi^1((P, s_1); e^1, h_Q) > \pi^1((Q, \cdot); e^1, h_Q). \tag{6.9}$$

Clearly, $d(\omega, \tilde{\omega}) = 1$.

Let $\tilde{h} = [((n - 1, 1, 1, 0), (1, 0)), ((n, 0, 0, 0), (0, 0))]$ denote the history induced from $h_Q$ when the individual for whom (6.9) applies switches to strategy $(P, s_1)$. From $\tilde{\omega}$, a transition to a state where the pattern of expectation of *every* agent of population 1 satisfies

$$e^1(s_1 \mid \tilde{h}) = 1$$

results merely from application of postulate SE under *no* mutation. From this point, if all agents of population 1 can simultaneously adjust their strategy (and do *not* mutate), the ensuing state $\breve{\omega} = \left[ \breve{h}, \left(\breve{z}^i(\psi^i)\right)_{\psi^i, i} \right]$ will satisfy

$$\sum_{\psi^1 \in \Psi_P^1} \breve{z}^1(\psi^1) = \sum_{\psi^2 \in \Psi_P^2} \breve{z}^2(\psi^2) = n.$$

If no further mutation occurs, the ensuing state $\hat{\omega} \in \Upsilon_P$. Therefore, the path $y = \{(\omega, \tilde{\omega}), (\tilde{\omega}, \breve{\omega}), (\breve{\omega}, \hat{\omega})\}$ links $\omega$ to a state in $\Upsilon_P$, as required. Moreover, $c(y) = 1$, which completes the proof of the Lemma.

**Lemma 21** *There exists some $\tilde{n} \in \mathcal{N}$ such that if $n \geq \tilde{n}$, $\forall \omega \in \Upsilon_Q$, $\forall \hat{\omega} \in \Upsilon_P$, and any path $y$ linking $\hat{\omega}$ to $\omega$, $c(y) > 1$.*

   *Proof:*  Let $\hat{\omega}$ be any state in $\Upsilon_P$. If just one mutation takes place from state $\hat{\omega}$, the second information set for type-1 players, $u_2^1$, is still reached in at least $\frac{n-1}{n}$ of the encounters (in fact, in *every* encounter if the mutation is on an agent of population 2). If $n$ is large enough, it is clear that the ensuing expectation pattern resulting from the application of SE must still make it optimal for every agent to choose strategy $s_1$ next period, thus leading back to a state belonging to $\Upsilon_P$. Thus, one mutation is *not* enough to trigger a transition to the component $\Upsilon_Q$. This completes the proof of both the Lemma and the Proposition.  ∎

In contrast with previous results, drift (more specifically, "expectational drift") is the key phenomenon at work in the present scenario. It crucially underlies the selection of the efficient co-ordination equilibrium, allowing the process to overcome the risk-dominance considerations which proved decisive in the context of Theorem 22 above.

The fact that, in the present multistage context, *not* all information sets need to be reached every period is the main feature which permits drift to have such an important effect. In particular, it is clear that neither the single information set of player 2 nor the second one of player 1 will be reached if *every* player of this latter type decides to eschew the co-ordination game altogether by choosing the outside option $Q$. But then, if this happens, players' expectations on the opponents' behaviour *at these information sets* may drift undisciplined by selection (i.e. updating) forces. In particular, a suitable concatenation of one-step mutations may by itself lead to a state where every player of population 2 believes that, if any of his opponents were to decide to play the co-ordination subgame, he would choose $s_1$. From such a situation (which is stationary for the mutation-free dynamics), only one further mutation is required to make *some* type-1 player expect that he will be responded to with $s_1$ if he decides to enter the co-ordination subgame by choosing $P$. If this mutation indeed occurs, those expectations are fully confirmed, this event immediately "locking in" all other players' expectations on the efficient path of play. Subsequently, a complete transition is triggered towards the efficient outcome (at least with some positive probability) without the need of any further mutation.

Once the population is playing the efficient co-ordination equilibrium, *all* information sets $(u_1^1, u_2^1,$ and $u^2)$ are being visited every period. Therefore, there is no range for analogous drift to materialize, since all conditional expectations are permanently being determined by currently observed (and confirming) evidence. In this case, a transition away from the efficient path of play requires a certain number of *simultaneous* mutations, exactly as in the pure co-ordination context studied in Subsection 6.4.1. When the population is large and $\epsilon$ small, these simultaneous mutations give rise to an event that is much more unlikely than the drift-based transition outlined above.

Interestingly enough, the evolutionary considerations just described end up selecting the only equilibrium of the game which is consistent with "rational" forward induction, as considered by classical Game Theory in a variety of alternative formulations (see, for example, van Damme (1989)). As the reader will recall, similar ideas also played a key role in the example discussed in 4.2 within a continuous-time, deterministic set-up. In a sense, the potential for drift at untested information sets may be seen as a certain test of robustness on mutation-free stationary states which discards weakly dominated strategies. As is well-known (cf. Fudenberg and Tirole (1991: 460–67)), iterative elimination of weakly dominated strategies reflects considerations of forward induction, at least in simple multistage games.[13]

To provide some heuristic understanding of this idea, suppose that the system is at a state where the inefficient equilibrium is being played. Then, one may view evolutionary drift as occasionally "sending the signal" that, if a player of population 1 is willing to play $P$, it is because he expects to be responded to with $s_1$. (Equivalently, of course, this can be interpreted as indicating that type-1 players are *not* planning to choose the dominated $s_2$). Sometimes this same drift will lead the system to a state where this signal is suitable reacted to by type-2 players (i.e. they actually play $s_1$). In that case, a transition to the efficient equilibrium is carried out, thus exposing the weakness of the inefficient equilibrium for which the co-ordination subgame was not reached.

## 6.6   Dynamic Expectations

### *6.6.1   Introduction*

Given the dynamic nature of the context in which players interact, it seems natural to allow for the possibility that they may rely on genuinely dynamic considerations in shaping their expectations. Of course, this is not meant to suggest that they must be viewed as fully understanding the underlying dynamic process, thus being able to predict "optimally" its future evolution. More in consonance with the evolutionary paradigm, agents should rather be conceived of as striving to detect *some* regularities in previous history, then building upon them to shape their current expectations.

Once we decide to go beyond the clear-cut static rule of expectation formation (Postulate SE above), there seems to be little theoretical ground for postulating any alternative *particular* rule that reflects the desired dynamic considerations. Instead, the reasonable course to take seems to be the formulation of certain natural conditions that *any* admissible updating rule should satisfy, thus eschewing any specific formulation for it.

Subsection 6.6.2 below proposes two such conditions. The first one is a weak demand of intertemporal consistency. The second condition requires that only relevant precedents should matter. Their implications for the analysis of simple co-ordination games are discussed in Subsection 6.6.3. In contrast with our former analysis, the introduction of dynamic expectation rules will be seen to have drastic effects on the long-run predictions of the model. Most crucially, it transforms the evolutionary process from a mechanism of equilibrium selection

---

[13] See Nöldeke and Samuelson (1993) and Swinkels (1993) for further discussion of these issues. As mentioned, the framework of Nöldeke and Samuelson is very similar to that described here. Swinkels, however, studies deterministic evolutionary dynamics, similar in spirit to that studied by Ritzberger and Weibull (1995). He shows that any set satisfying a certain criterion of evolutionary stability must include a *hyper-stable set*, in the sense of Kohlberg and Mertens (1986). This set always satisfies a certain notion of forward induction.

into one displaying equilibrium volatility (i.e. all equilibria end up being played in the long run).

### 6.6.2    Admissible updating rules

As advanced, admissible updating rules will be required to satisfy the following two requirements:

(a)  Induced expectations have to be consistent with any sufficient long string of stationary evidence.

(b)  The revision of expectation patterns must be restricted to those contingencies for which observed history is relevant.

The formalization of each of these conditions is addressed in turn.

### 6.6.2.1    Consistency with long stationary evidence

Only a weak criterion of historical consistency will be considered here. It requires that if a certain *given* outcome has been *constantly* observed for a sufficiently long stretch of time in the past, the expectation pattern should predict that it will continue to repeat itself (of course, as long as no contradictory evidence is realized).[14] Formally, it may be described as follows:

**Constant Evidence (CE)** $\exists r \in \mathcal{N}$ such that $\forall \bar{h} = \left[ (\bar{x}_0^1, \bar{x}_0^2) \right]^2$, $\forall i = 1, 2,$
$\forall \{ e_v^i \}_{v=1,2,\ldots} \subset E^i,\ \left[ e_v^i = F^i(e_{v-1}^i, \bar{h}),\ v = 1, 2, \ldots \right] \Rightarrow$

$$
\left[
\begin{array}{ll}
\multicolumn{2}{c}{\forall v \geq r, \forall u \in U^j, \forall a \in A(u),} \\[2mm]
e_v^i(a \mid \bar{h}) = \dfrac{\bar{x}_0^j(a)}{\sum_{a' \in A(u)} \bar{x}_0^j(a')} & \text{if } \sum_{a' \in A(u)} \bar{x}_0^j(a') > 0 , \\[4mm]
e_v^i(\cdot \mid \tilde{h}) = e_{v-1}^i(\cdot \mid \tilde{h}) & \text{for all } \tilde{h} \neq \bar{h}.
\end{array}
\right]
$$

The former condition can be seen as a substantial relaxation of Postulate SE (cf. Section 6.3).[15] Unlike this former postulate, expectation updating is required to be static only when arbitrarily long evidence to that effect (i.e. confirming that the opponent's behaviour indeed remains static) has been observed. Otherwise, no requirement whatsoever is imposed by CE on the expectation-updating rule.

---

[14] Similar conditions of historical consistency have been often proposed in the learning literature. See e.g. Milgrom and Roberts (1991).

[15] Note that nothing is required on $e_v^i(a \mid \bar{h})$ if $a \in A(u)$ and $\sum_{a' \in A(u)} \bar{x}_0^j(a') = 0$. In line with SE, it could be demanded that, in this case, $e_v^i(a \mid \bar{h}) = e_{v-1}^i(a \mid \bar{h})$. However, since any such requirement is irrelevant for our future analysis (which will focus on simultaneous games), it is simply dispensed with.

### 6.6.2.2 *No revision after irrelevant evidence*

The second condition required from expectation-updating rules is of a quite different nature. It expresses the idea that agents should not revise their pattern of expectations for those contingencies for which the recently observed history is not relevant. Of course, the key question here is what histories are to be conceived as "relevant" for any particular contingency. The following criterion is proposed.

Let $h_t$ be the history observed up to (and including) some given $t$. Every player of population $i = 1, 2$ is assumed to rely on this history to update his current expectation pattern (recall Section 6.2). Two different sub-histories of $h_t$ need to be singled out. On the one hand, there is the *prior sub-history* $h_t^-$ which includes the outcomes in $h_t$ observed *before* the current period $t$. In our simple case, where histories only include two periods, we have $h_t^- = (x_{t-1}^1, x_{t-1}^2)$. On the other hand, there is what may be called the *recent sub-history* $h_t^+$. It includes all outcomes in $h_t$ except its very earliest one. Since in our simplified context this earliest outcome is merely given by $(x_{t-1}^1, x_{t-1}^2)$, one has $h_t^+ = (x_t^1, x_t^2)$.[16]

In updating the expectations which prevailed at $t$, the implicit assumption is that every player will attempt to identify some regularities in the prior history $h_t^-$ and associate them with the ensuing outcome observed at $t$, $(x_t^1, x_t^2)$. Thus, along these lines, only when a given $\hat{h} \in H$ has a sub-history $\hat{h}^+$ which matches $h_t^-$ can history $h_t$ be taken as a relevant precedent to assess the outcome which will follow $\hat{h}$. For all other $\tilde{h} \in H$ such that $\tilde{h}^+ \neq h_t^-$, the associated conditional expectations should remain unaffected by the observation of the ("irrelevant") history $h_t$. This idea may be formalized as follows.

**Relevant Evidence (RE)** $\forall i = 1, 2, \ \forall e^i \in E^i, \ \forall h \in H$,

$$\hat{e}^i = F^i(e^i, h) \Rightarrow \forall \tilde{h} \in H : \tilde{h}^+ \neq h^-, \ \hat{e}^i(\cdot \mid \tilde{h}) = e^i(\cdot \mid \tilde{h}).$$

In a heuristic sense, the previous condition shares some of the spirit of Postulate SE. As in it, expectations associated with unobserved realizations (there, information sets; here, prior histories) are required to be left unchanged by the updating process.

### 6.6.3 *Equilibrium volatility*

For a sharper contrast with previous results, we focus again on the co-ordination set-up analysed in Subsection 6.4.1. Thus, individuals are paired to play a

---

[16] Thus, recent sub-history and "current" outcome coincide. In general, of course, if histories included more than two periods, this would not be the case. Specifically, if they consisted of the last three periods, the *recent history* would include both the current and previous observations. It should be clear how the approach and motivation discussed here are fully extendable to histories of any arbitrary – but finite – length.

symmetric $2 \times 2$-game, with pay-off structure as described in Table 15 and

$$a > c, \ b > d.$$

Under no further restrictions on the game pay-offs, the next result establishes that the evolutionary process will lead to what could be labelled as long-run "equilibrium volatility". That is, the two equilibria of the co-ordination game (but *only* equilibria) will be played, a.s., *some* significant fraction of the time in the long run.

**Theorem 23 (Vega-Redondo (1995))** *Assume CE and RE. Then, for each* $q = 1, 2,$ *there exists some* $\tilde{\omega} = [\tilde{h}, (\tilde{z}^i(\psi^i))_{\psi^i, i}] \in \Omega$ *such that* $\lim_{\epsilon \to 0} \mu_\epsilon(\tilde{\omega}) > 0$ *and*

$$\tilde{z}^i(s^i, e^i) > 0 \ (i = 1, 2) \Rightarrow \left[ s^i = s_q \ \wedge \ e^i(s_q \mid \tilde{h}) = 1 \right]. \tag{6.10}$$

*Conversely, any state* $\tilde{\omega} \in \Omega$ *which displays* $\lim_{\epsilon \to 0} \mu_\epsilon(\tilde{\omega}) > 0$ *satisfies (6.10) for some* $q = 1, 2.$

**Proof.** Define $h_q$, $\Psi_q^i$, $\Upsilon_q$ as in the proof of Theorem 22 (see (6.4) and (6.5)). First, the absorbing sets of the mutation-free dynamics are narrowed down as follows.[17]

**Lemma 22** *Let* $\omega \in \Omega$ *belong to an absorbing set of the mutation-free dynamics. Then,* $\omega \in \Upsilon_1 \cup \Upsilon_2$.

*Proof:*  Consider any $\omega^0 = \left[ h, (z(\psi^i))_{\psi^i, i} \right] \in \Omega$ prevailing at some given $t$. It is argued that there is a finite chain of transitions $\{(\omega^0, \omega^1), (\omega^1, \omega^2), ..., (\omega^{m-1}, \omega^m)\}$ which has positive probability (bounded above zero) and such that $\omega^m \in \Upsilon_q$ for some $q = 1, 2$. Moreover, such $\omega^m$ is a stationary state of the mutation-free dynamics.

Suppose first that every individual of both populations chooses some given strategy $s_q$ at $\omega^0$. That is,

$$z^{i0}(s^i, e^i) > 0 \ (i = 1, 2) \Rightarrow s^i = s_q . \tag{6.11}$$

Then, one just needs to choose $m > r$ (where $r$ is the pre-established number of steps contemplated by Postulate CE) and assume that no individual is able to adjust his action for the ensuing $m$ periods. Then, by CE, one has

$$z^{im}(s^i, e^i) > 0 \ (i = 1, 2) \Rightarrow \left[ s^i = s_q \ \wedge \ e^i(s_q \mid h_q) = 1 \right].$$

---

[17] If, as in Vega-Redondo (1995), one required the condition that all admissible expectation patterns $e^i \in E^i$ satisfy

$$\forall h, \hat{h} \in H, \ h^+ = \hat{h}^+ \Rightarrow e^i(\cdot \mid h) = e^i(\cdot \mid \hat{h}),$$

i.e. expectations only depend on the latest observed outcome of prior history, then every state in $\Upsilon_1 \cup \Upsilon_2$ would be stationary in the mutation-free dynamics. In that case, the statement of Lemma 22 and that of the Theorem could be expressed in terms of a single necessity and sufficiency requirement.

Since the history prevailing at state $\omega^m$, $h^{m-1}$, is obviously equal to $h_q$, and so it happens for all other previous $\omega^v$, $v = 2, 3, ..., m-1$, it follows from Postulate CE both that $\omega^m \in \Upsilon_q$ and that this state is stationary for the mutation-free dynamics. Moreover, given the postulated formulation for strategy adjustment, such a chain has an *ex ante* probability not less than $\left[(1-p)^{2n}\right]^m$.

Suppose now that (6.11) is not satisfied at $\omega^0$. Then, one needs to consider a longer chain of transitions $\{(\omega^0, \omega^1), (\omega^1, \omega^2), ..., (\omega^{m-1}, \omega^m)\}$ as follows.

Divide this chain into five subchains of respective lengths $m_1, m_2, ..., m_5$, $m \equiv \sum_{j=1}^5 m_i$. In the first subchain, with length $m_1 \geq r$, all individuals remain throughout with their initial strategy. As before, this subchain has prior probability not less than $\left[(1-p)^{2n}\right]^{m_1}$. At the end of it, due to Postulate CE, all individuals of any given population will share the same respective expectations contingent on the relevant history. That is

$$\left[z^{im_1}(s^i, \mathbf{e}^i) > 0, \; z^{im_1}(\hat{s}^i, \hat{\mathbf{e}}^i) > 0 \; (i = 1, 2)\right] \Rightarrow \; e^i(\cdot \mid \bar{h}) = \hat{e}^i(\cdot \mid \bar{h}),$$

where $\bar{h}$ is the constant history induced by the initial profile of strategies.

The second subchain consists of just one transition. In it, all individuals of one of the populations, say population 1, switch to an identical strategy $s_q$ which is optimal given their identical relevant beliefs. This transition has probability at least equal to $\left(\frac{1}{2}p\right)^n (1-p)^n$, due to the assumption that all strategies in the current set of optimal best responses are chosen with equal probability.

The third subchain again has length $m_3 \geq r$. It is analogous to the first subchain in that all individuals are assumed to remain with their initial strategy, at the end of which every one of them is sharing the same expectation contingent on the prevailing history. In particular, all individuals of population 2 believe with probability one that *every* member of population 1 will play strategy $s_q$, as specified above. The prior probability of this subchain is $\left[(1-p)^{2n}\right]^{m_3}$.

Given these beliefs, the fourth subchain, again composed of just one transition, involves all players of population 2 switching to strategy $s_q$. This transition has probability not less than $p^n(1-p)^n$.

From this point onwards, one just needs to replicate the chain considered in the first part of the argument (under the assumption that all individuals of both populations keep playing the same strategy). Let $m_5$ be the length of this last chain. Its prior probability is at least $\left[(1-p)^{2n}\right]^{m_5}$. At the end of it, the state $\omega^m \in \Upsilon_q$, as explained above. Moreover, by Postulate CE, one may also conclude that $\omega^m$ is a stationary state for the mutation-free dynamics. Since, combining the former considerations, it follows that the prior probability of the composite chain can be bounded above zero, independently of the initial state $\omega^0$, the proof of the Lemma is complete.

Let $\mathcal{A}_q$, $q = 1, 2$, denote the collection of states in $\Upsilon_q$ which, viewed as singletons, are absorbing sets of the mutation-free dynamics. Obviously, each $\mathcal{A}_q \neq \emptyset$. To complete the proof of the Theorem, two further steps are required. The first one is based on the observation that, within each $\mathcal{A}_q$, states only differ in those players' expectations which are contingent on histories *not* observed within it. Thus, relying on an argument identical to that of Lemma 19 above, the following analogous result can be shown.

**Lemma 23** *Let* $\omega, \omega' \in \mathcal{A}_q$, *for some given* $q = 1, 2$. *Then,*

$$\min_{Y \in \mathcal{Y}_\omega} c(Y) = \min_{Y \in \mathcal{Y}_{\omega'}} c(Y).$$

And then, in view of the preceding result, the proof of the Theorem follows from the following final Lemma.

**Lemma 24** *Let* $\omega \in \mathcal{A}_q$ *for some given* $q = 1, 2$. *Then there exists some* $\omega' \in \mathcal{A}_{q'}$, $q' = 3 - q$, *such that*

$$\min_{Y \in \mathcal{Y}_{\omega'}} c(Y) \leq \min_{Y \in \mathcal{Y}_\omega} c(Y).$$

*Proof:*   Consider any arbitrary $\tilde{\omega} \in \mathcal{A}_q$. By the definition of this set and postulate RE, there is some $\hat{\omega} = \left[ h_q, \left( \hat{z}(\psi^i) \right)_{\psi^i, i} \right] \in \mathcal{A}_q$ such that:

$$\hat{z}^i(s^i, e^i) > 0 \ (i = 1, 2) \Rightarrow \forall h \in H \text{ s.t. } h^+ \neq h_q^+, \ e^i(s_{q'} \mid h) = 1, \ q' = 3 - q. \tag{6.12}$$

Let $\hat{Y}$ be the $\hat{\omega}$-tree which achieves the minimum cost in $\mathcal{Y}_{\hat{\omega}}$. From Lemma 23,

$$\min_{Y \in \mathcal{Y}_{\hat{\omega}}} c(Y) = c(\hat{Y}).$$

Now, construct a path $y$ joining $\hat{\omega}$ to some state $\omega' \in \mathcal{A}_{q'}$ ($q' = 3 - q$) as follows. First, consider the state derived from $\hat{\omega}$ when a single player of, say, population 1 mutates by choosing strategy $s_{q'}$. Let $\omega^1 = [h^1, \left( z^{1i}(\psi^i) \right)_{\psi^i, i}]$ denote the state induced by this mutation. By construction $d(\hat{\omega}, \omega^1) = 1$. Consider then the state $\omega^2 = [h^2, \left( z^{2i}(\psi^i) \right)_{\psi^i, i}]$ derived from $\omega^1$ when every player of each population adjusts his strategy and experiences *no* mutation (therefore $d(\omega^1, \omega^2) = 0$). From (6.12) and Postulate RE, it follows that

$$z^{2i}(s^i, e^i) > 0 \ (i = 1, 2) \Rightarrow e^i(s_{q'} \mid h^1) = 1.$$

Therefore, state $\omega^2$ also has

$$z^{2i}(s^i, e^i) > 0 \ (i = 1, 2) \Rightarrow s^i = s_{q'}.$$

Suppose now that players remain with their strategies fixed for $m$ periods and let $\omega^3, \omega^4, ..., \omega^{m+2}$ be the corresponding states. Of course, one has

$$d(\omega^v, \omega^{v+1}) = 0, \quad v = 2, ..., m+1,$$

due to the assumed positive (and independent) probability of inertia affecting all players each period. If $m > r$, where $r$ is the pre-established parameter specified by Postulate CE, it follows that $\omega^{m+2} = \omega' \in \mathcal{A}_{q'}$, as desired.

Finally, it is argued that there is an $\omega'$-tree $Y'$ which satisfies $c(Y') \leq c(\hat{Y})$. The argument is constructive. First, consider the $m+3$ (incomplete) sub-trees associated with each of the states in the path $y$, obtained by deleting the arrows starting in each of them in the (full) tree $\hat{Y}$. Then, add the arrows included in the path $y$. This gives rise to an $\omega'$-tree $Y'$. The total cost of this operation is $c(y) = 1$. On the other hand, note that there has been a decrease of at least one cost unit by eliminating the arrow starting at $\omega'$ in tree $\hat{Y}$. (This follows from the fact that $\omega'$ is a stationary point of the mutation-free dynamics and, therefore, it cannot be exited without mutation.) Overall, it may be concluded that $c(Y') \leq c(\hat{Y})$, as claimed. This completes the proof of the Lemma and, thus, of the Theorem. ∎

The previous result tempers any excessive hopes one might have entertained on the potential of evolutionary processes as equilibrium-selection mechanisms. In a sense, it reinforces the widely accepted view that equilibrium selection is an issue which cannot generally be addressed in the abstract, i.e. without some reference to the particular context of application.

If players are relatively sophisticated and can view their context as what it indeed is, i.e. a dynamic one, expectational drift at untested contingencies may lead to (dynamic) expectations which interpret any change in the *status quo* as a "transition signal". This, in the end, brings about equilibrium volatility: the impossibility of anchoring the system at a particular equilibrium. Such a conclusion is underscored by the fact that, in the context of Theorem 23, no restrictions are imposed on the game pay-offs other than assuming that there is some selection issue to be concerned with (that is, there are two different equilibria in which players can be co-ordinated). Thus, its conclusion applies even if one of the equilibria dominates the other one, both in the Pareto and the risk-dominance sense.[18]

This raises the question of how much sophistication is it reasonable to assume on the part of the players. One should always bear in mind that, as stressed by Mailath (1992), modelling is a "metaphorical" activity, which is never to be interpreted literally. From this perspective, the assumption that players' sophistication is quite limited (even as limited as postulated implicitly in Chapter 5) might not be a bad modelling assumption at all. In the real world (of which our models are supposed to be a schematic sketch) players *do* actually behave in a very simple way *relative* to the complexity of the environment. When these key relative considerations are contemplated within an unavoidably simple (if manageable) model, quite simple rules of behaviour may well appear to be the most interesting ones.

This, of course, does not make it less theoretically interesting to explore the sensitivity of our approach to the consideration of different degrees of players'

---

[18] In fact, as discussed in Vega-Redondo (1995), volatility will prevail in any arbitrarily large co-ordination game, among all of its pure-strategy equilibria.

sophistication. In a sense, this is partly what much of this chapter has been concerned with. Now, the next natural step is undertaken: the introduction of "sophistication" as an endogenous component of the model. A preliminary, simply suggestive attempt in this direction closes the chapter.

## 6.7    On the Evolution of Sophistication

### *6.7.1    Introduction*

Throughout this and previous chapters, we have explored how a (stochastic) evolutionary approach to the analysis of games is affected by the specification of different details on the underlying environment, e.g. the number of matching rounds taking place every period (cf. Subsections 5.4.1 and 5.4.2) or the particular nature of the interaction pattern (cf. Subsections 5.6.1 and 5.6.2). But, arguably, the most interesting variations explored pertain to the alternative behavioural paradigms which are, more or less explicitly, embodied by the different frameworks considered.

Thus, in Chapter 5, no expectation-based considerations were involved and, in line with the more traditional evolutionary approach, the evolution of a certain strategy was linked to its relative (average) performance. A natural interpretation of this formulation is that players are mere *imitators*. That is, they simply observe the average pay-off earned by the different strategies and decide to mimic that one which is highest.[19] Such an interpretation presumes no knowledge whatsoever on any agent's part about the structure of the game he is playing, not even about the details which affect him directly (for example, his own pay-off structure). This will be one of the levels of sophistication (the lowest one) which will be considered below. It postulates that players are (or at least behave as if they were) pure imitators, along the lines described.

The second higher level of sophistication considered will assume players can form expectations on the evolution of the process, but only *static* ones. Thus, as specified in Section 6.3, it will postulate players who observe the previous outcome and simply predict that other agents will keep playing as observed. Then, on the basis of such predictions, they simply choose an optimal response. Note, of course, that this formulation implicitly embodies the assumption that players are at least aware of their own best-response correspondence. For the sake of a better label, agents of this kind will be described as *myopic optimizers*.

Finally, the third (highest) level to be considered involves players whose expectations are shaped dynamically (in the manner described in the previous section), and respond optimally to whatever beliefs their current expectation patterns and prior history induce. In an intuitive sense, these players exhibit

---

[19] Recall Subsection 4.3.3, where an analogous specification was suggested within a continuous-time and deterministic evolutionary framework

a degree of sophistication much higher than the two previous ones. Not only must they know their own pay-off structure, but they must also be able to "keep track" of a whole pattern of conditional expectations which are updated as the process evolves.[20] Such players will be called *dynamic optimizers*.

The theoretical exercise to be conducted may be outlined as follows. Consider a context where, in principle, players of different degrees of sophistication (i.e. different types) may coexist. (Here, the term *type* will be reserved to identify the player's degree of sophistication. No confusion should arise with our previous use of it to specify the population to which a player belongs.) As the process unfolds, each individual adjusts his behaviour and other characteristics as described above, depending on his respective type. Furthermore, with some given probability, every one of them also has the opportunity of changing his type. In that case, he is assumed to revise it only if some other type has obtained a higher average pay-off in recent ("recorded") history. Again, perturbing the process by sporadic and independent mutation (which affects both characteristics *and* types), the question then posed is the usual one: What is the long-run behaviour of the process as the mutation probability becomes small?

As it turns out, the answer to the specific question of what *equilibrium behaviour* is selected depends crucially on the range of types admitted as possible. This will be seen to reinforce previous considerations, to the effect that evolutionary mechanisms are more effective the less sophisticated players are assumed (or allowed) to be. On the other dimension of the process, i.e. selection of types, the effectiveness of the process in narrowing down the set of long-run types will be shown independent (perhaps disappointingly so) of the range of types under consideration. Specifically, an evolutionary process of the kind proposed will prove unable to discard any type as a "viable" candidate in the long run, irrespective of its relative sophistication.

### 6.7.2    The model

As before, individuals of two populations (with equal size $n$) are assumed randomly matched in a given number of rounds every period in order to play a $2 \times 2$-game of co-ordination with the pay-off structure described in Table 15. For every period $t$, the state of the process includes a specification of how many individuals display a certain characteristic. Now, however, the space of characteristics not only includes the strategy and the expectation held by a given player but also his type. Let $\mathcal{T}^i \subset \{\iota, \sigma, \delta\}$ be the type space under consideration for population $i$, where $\iota$ stands for "imitator", $\sigma$ for "static optimizer",

---

[20] It is true that, within the general framework proposed in this chapter, players are always assumed to hold a *complete* pattern of expectations, even if they update their expectations in a static manner. In this latter case, however, all of its components except the one associated with the prevailing history do not have any effect on future behaviour and, therefore, may be fully ignored by these agents. In this sense, myopic optimizers are less sophisticated than dynamic optimizers.

and $\delta$ for "dynamic optimizer". Adhering to previous notational conventions for the sake of simplicity, the space of characteristics of population $i$ is given by $\Psi^i = S^i \times E^i \times T^i$. (Of course, for any individual who is an imitator, his expectation component is irrelevant. In fact, it may as well be assumed that he holds *no* expectations, just attributing to him some constant "dummy" element of his respective $E^i$.)

The state of the process at $t$ must also include the current *history*. Again, this object must be somewhat more complicated than in the original framework. It must certainly include the outcomes realized in the current and previous period $\left[(x_0^1, x_0^2), (x_1^1, x_1^2)\right]$, where the term "outcome" is interpreted as above, i.e. a specification of how many individuals have been observed to choose each of the possible actions. However, there is an additional piece of information which should also be included in the current state of the system if its description is to be sufficient. Specifically, the average pay-offs earned by each type and every action must also be specified. The first bit of this information (average pay-offs of types) is needed to formulate the type-adjustment dynamics outlined above. On the other hand, information on the average pay-offs earned by each strategy is required by any imitator who might be in a position to revise his strategy.[21]

Therefore, histories $h \in H$ (again, adopting prior notation) must be vectors of the following form $h = \left[(x_k^1, \pi_k^1), (x_k^2, \pi_k^2)\right]_{k=0,1}$, where each $x_k^i$ has the former interpretation and every $\pi_k^i$ is a five-dimensional vector which, for each population $i = 1, 2$, specifies the average pay-off $\pi_k^i(\cdot)$ earned by each type $\tau^i \in T^i$ and every strategy $s_q \in S^i$ in the current ($k = 0$) and previous ($k = 1$) periods. If one of the strategies or types is *not* represented in the population, it is formally assigned a small enough pay-off value (say $-\infty$), thus guaranteeing that it will never be adopted through imitation alone – recall Chapter 5, where a similar assumption was made.

The mutation-free dynamics of the process may be then described as follows. Every period $t$, each individual of population $i = 1, 2$ carries out the following sequential operations. (In order to avoid cumbersome notation, some formalities are dispensed with. The reader familiar with the previous developments should have no trouble filling in the formal details.)

(i) *Type adjustment*

With probability $\theta \in (0, 1)$, the agent receives the opportunity of adjusting his prior type $\tau^i \in T^i$. In that case, he will change to a new type if, and only if, there is another type that, on average over past history $h_{t-1}$, has earned a higher pay-off than his own. If several such exist, every one of them is chosen with equal probability.[22]

---

[21] Of course, information on the average pay-offs earned by each strategy would not be needed if, as in Subsection 5.4.2, the focus were on a context with small matching noise (that is, if we made the number of rounds grow to infinity before the limit on $\epsilon$ is undertaken). However, in this context, the behaviour of imitators and myopic optimizers is essentially equivalent, thus rendering redundant the inclusion of both types. This is the reason why the number of rounds is maintained fixed in the present analysis.

[22] Different formulations could be postulated as well e.g. the average pay-offs earned by each

If the new type is $\delta$ (i.e. the agent becomes a dynamic optimizer), the question arises as to what expectation pattern the agent should hold after the switch. If any expectation pattern were allowed at this point, type adjustment could act as a disguised mutation. To prevent this, a number of alternative possibilities could be considered. For the sake of concreteness, it will be simply assumed that the new expectations are chosen to coincide with those previously held by one of the individuals with dynamic expectations.[23]

### (ii) *Expectation updating*

Depending on whether the player's type is relevant in this respect (i.e. it is either $\sigma$ or $\delta$), his expectations are adjusted accordingly. That is, they are updated as indicated by Postulate SE if his type is $\sigma$, or some given updating rule $F^i(\cdot)$ which satisfies Postulates CE and RE if his type is $\delta$.

### (iii) *Strategy adjustment*

The player enjoys a common and independent probability $p \in (0, 1)$ of adjusting his strategy. If he receives a strategy-revision opportunity, he chooses the new strategy as dictated by his current type. Specifically, if he is an imitator, he chooses the strategy which earned the highest average pay-off over history $h_{t-1}$.[24] Otherwise, he chooses a strategy which is a best response to his current beliefs, as induced by prior history and his current pattern of expectations.

Items (i) to (iii) define the mutation-free dynamics of the process. As usual, this dynamics will be perturbed by introducing mutation. That is, prior to play being conducted every period, each individual is subject to a common and independent probability $\epsilon > 0$ of mutation. If this event happens to materialize for any given individual, his new characteristic is chosen from the respective $\Psi^i$ according to a given probability distribution with full support.

The combination of these components yields an ergodic stochastic process on $\Omega$ with unique invariant distribution $\mu_\epsilon \in \Delta(\Omega)$. Its long-run behaviour is studied below (for small $\epsilon$) in two alternative scenarios. First, the analysis focuses on a context where the "sophistication range" is relatively "narrow". This is taken to mean that the set of available types is simply $T^i = \{\iota, \sigma\}$ for each $i = 1, 2$. The second scenario, with a "wide sophistication range", has $T^i = \{\iota, \sigma, \delta\}$ for each $i = 1, 2$.

---

type in the earlier past could be assigned a lower weight (even one equal to zero). Or the *status quo* type could be assigned no priority in the adjustment rule. All of these variations would yield equivalent implications.

[23] Note, of course, that at least one such individual must exist if this type is worth "imitating". (Otherwise, this type's average pay-off would be $-\infty$, as postulated above.) Another possibility would be to assume that (under the implicit assumption that the player has no prior relevant information) *initial* expectations are just static for every conceivable history.

[24] To match exactly the framework discussed in Ch. 5, only the pay-offs earned in the previous period should be involved. As explained above for the adjustment on types, variations on these matters are inessential.

### 6.7.3    Narrow sophistication range

As explained, we assume first that $T^i = T_0 \equiv \{\iota, \sigma\}$ for each $i = 1, 2$. In this case, the next result shows that the evolutionary process selects the efficient equilibrium in the long run.

**Theorem 24** *Assume $a > b$, and $T^i = T_0$ for each $i = 1, 2$. Then there exists some $\tilde{n} \in \mathcal{N}$ such that if $n \geq \tilde{n}$, any $\tilde{\omega} \in \Omega$ with $\lim_{\epsilon \to 0} \mu_\epsilon(\tilde{\omega}) > 0$ satisfies $\tilde{x}_k^i(s_1) = n$ for each $i = 1, 2$, $k = 0, 1$.*

**Proof** (sketch). The proof is just outlined here, since it relies on ideas that have already been explained in detail at different points in this chapter. First, it can be verified that the absorbing sets of the mutation-free dynamics are as indicated by the following Lemma.

**Lemma 25** *Let $\tilde{\omega} = \left[\tilde{h}, \left(\tilde{z}^i(\psi^i)\right)_{\psi^i, i}\right] \in \Omega$ belong to some absorbing set of the mutation-free dynamics. Then, $\tilde{\omega}$ is stationary for this dynamics and there exists some $q = 1, 2$ such that*

$$\tilde{z}^i(s^i, \mathbf{e}^i, \tau^i) > 0 \quad (i = 1, 2) \Rightarrow s^i = s_q.$$

The proof of this Lemma is essentially a combination of the arguments used in the proofs of Theorems 19, 22, and 23. From it, one may conclude that those states which, as singletons, are absorbing sets of the mutation-free dynamics can be partitioned into two classes, $\mathcal{A}_q^*$, $q = 1, 2$, analogous to those defined in the proof of Theorem 23. In each of them, the same strategy $s_q$ is respectively being played by *all* individuals in every state.

The following instrumental results also apply to these sets.

**Lemma 26** *Let $\omega, \omega' \in \mathcal{A}_q^*$, for some given $q = 1, 2$. Then,*

$$\min_{Y \in \mathcal{Y}_\omega} c(Y) = \min_{Y \in \mathcal{Y}_{\omega'}} c(Y).$$

**Lemma 27** *For each $q = 1, 2$, and every given $\hat{\tau} \in T_0$, there exists some $\tilde{\omega} = \left[\tilde{h}, \left(\tilde{z}^i(\psi^i)\right)_{\psi^i, i}\right] \in \mathcal{A}_q^*$ such that*

$$\tilde{z}^i(s^i, \mathbf{e}^i, \tau^i) > 0 \quad (i = 1, 2) \Rightarrow \tau^i = \hat{\tau}. \tag{6.13}$$

The proof of Lemma 26 is the counterpart of Lemma 23 above and is based on the same underlying fact, i.e. all states in any given $\mathcal{A}_q^*$ can be connected through one-step mutations. This fact follows from the following two observations:

(i)  Expectation patterns in any given $\mathcal{A}_q^*$ can only differ for contingencies which are not observed within this set (recall the argument of Lemma 19).

i) Each of the two possible types in $\mathcal{T}_0 = \{\iota, \sigma\}$ leads to the same strategy and receives the same pay-off within any given $\mathcal{A}_q^*$. Thus, any mutation of one agent from one type to the alternative one (keeping his action unchanged) will make the system remain in this set.

The latter point can be directly used to prove Lemma 27. Based on this lemma, the rest of the proof may be outlined as follows. Consider any given state in $\mathcal{A}_2^*$ where (6.13) is satisfied for $\hat{\tau} = \iota$. From such a state, a transition can be triggered towards $\mathcal{A}_1^*$ with a number of mutations, say $k$, that is independent of population size. The argument is as in Lemma 5 in Chapter 5. On the other hand, from Lemmas 6 and 14, it should be clear that the number of mutations required to implement the converse transition from any $\omega \in \mathcal{A}_1^*$ to some $\omega' \in \mathcal{A}_2^*$ is bounded below by some given proportion of $n$, the common size of both populations. Thus, it is always larger than $k$ if $n$ is large enough. These two facts, combined with Lemma 26, yield the desired conclusion by relying on the usual "tree-pruning operations" repeatedly used in this chapter. ∎

Theorem 24 addresses explicitly only one of the dimensions of the process that interests us here: the equilibrium behaviour observed in the long run. It establishes that if imitators and myopic optimizers are the only possible types, the former play the crucial role in determining the long-run behaviour of the system. Specifically, the populations end up spending most of the time playing the efficient equilibrium, the outcome which was shown to prevail when *all* agents are always imitators.

About the other interesting dimension of the process, the evolution of types, the key issue concerns what level(s) of sophistication is selected in the long run. To address it formally, let $\lambda_\epsilon \in \Delta(\mathcal{T})$ denote the marginal density on the type space induced by the corresponding distribution $\mu_\epsilon$ on $\Omega$. Directly from Lemma 27, one obtains the following corollary:

**Corollary 6** *Under the conditions of Theorem 24,* $\lim_{\epsilon \to 0} \lambda_\epsilon(\tau) > 0$ *for each* $\tau \in \mathcal{T}_0$.

Thus, as advanced, the evolutionary process does not select, uniquely, between any of the two possible types considered here (i.e. between the two alternative levels of sophistication considered). A discussion and interpretation of this state of affairs is postponed to Subsection 6.7.5 below. Next, we turn to analysing how these matters are affected if the type space is enlarged to include dynamic optimizers as well.

### 6.7.4    *Wide sophistication range*

Suppose now that the type space equals $T_1 \equiv \{\iota, \sigma, \delta\}$ for both populations. In this case, the next result establishes that the evolutionary process will lead to the kind of equilibrium volatility obtained in Subsection 6.6.3, when only dynamic optimizers were allowed.

**Theorem 25** *Let $T^i = T_1$ for each $i = 1, 2$. For each $q = 1, 2$, there exists some $\tilde{\omega} = [\tilde{h}, (\tilde{z}^i(\psi^i))_{\psi^i, i}] \in \Omega$ with $\lim_{\epsilon \to 0} \mu_\epsilon(\tilde{\omega}) > 0$ such that*

$$\tilde{z}^i(s^i, e^i, \tau^i) > 0 \quad (i = 1, 2) \Rightarrow s^i = s_q. \tag{6.14}$$

*Conversely, every state $\tilde{\omega} \in \Omega$ which displays $\lim_{\epsilon \to 0} \mu_\epsilon(\tilde{\omega}) > 0$ also satisfies (6.14) for some $q = 1, 2$.*

**Proof** (sketch).   Lemmata 25 and 26, stated above for Theorem 24, apply in the present context without modification. On the other hand, the analogue of Lemma 27 can be formulated as follows.

**Lemma 28** *For each $q = 1, 2$, and every given $\hat{\tau} \in T_1$, there exists some $\tilde{\omega} = \left[\tilde{h}, (\tilde{z}^i(\psi^i))_{\psi^i, i}\right] \in \mathcal{A}_q^*$ such that*

$$\tilde{z}^i(s^i, e^i, \tau^i) > 0 \quad (i = 1, 2) \Rightarrow \tau^i = \hat{\tau}. \tag{6.15}$$

The proof may then be completed as follows. Consider any given $q = 1, 2$ and let $\hat{\omega} \in \mathcal{A}_q^*$ be a state where (6.15) is satisfied for $\hat{\tau} = \delta$. Moreover, in analogy with (6.12), assume that $\hat{\omega}$ satisfies:

$$\hat{z}^i(s^i, e^i, \tau^i) > 0 \; (i = 1, 2) \Rightarrow$$
$$\forall h \in H \text{ s.t. } h^+ \neq h_q^+, \; e^i(s_{q'} \mid h) = 1, \; q' = 3 - q.$$

Such a state exists, due to considerations explained in the proofs of Theorems 23 and 24. One may then rely on a familiar line of argument to conclude that there is a path $y$ joining $\hat{\omega}$ to some state $\omega' \in \mathcal{A}_{q'}^*$, $q' \neq q$, with cost $c(y) = 1$. (Note that the type-profile prevailing at $\hat{\omega}$, i.e. everyone being a dynamic optimizer, can be held fixed through any finite chain of transitions with positive probability. Since this construction is independent of the particular $q$ and $q'$ chosen (that is, it is fully symmetric between $\mathcal{A}_1^*$ and $\mathcal{A}_2^*$), it follows that the transition across these sets is equally costly in either direction. Relying on the usual graph-theoretic techniques, this leads to the desired conclusion.   ∎

In parallel with Corollary 6, Lemma 28 leads to the following result pertaining to the long-run marginal distribution on the type space.

**Corollary 7** *Under the conditions of Theorem 25, $\lim_{\epsilon \to 0} \lambda_\epsilon(\tau) > 0$ for each $\tau \in T_1$.*

### 6.7.5   Discussion

The contrasting conclusions of Theorems 24 and 25 support the heuristic notion that evolutionary processes will be the most effective in singling out efficient behaviour when the sophistication of players is not too high. To a certain extent, this idea has already emerged from our previous analysis (cf. Theorems 19, 22, and 23). The present approach substantially reinforces it by allowing the degree of sophistication of players to be *endogenously* determined by the evolutionary system. In this context, the former results can be interpreted as specifying a certain sophistication threshold below which efficient performance still results in the long run.

Thus, Theorem 24 establishes that the long-run behaviour of the system will be concentrated on the efficient equilibrium if, despite the fact that agents have the potential of behaving as static optimizers, they *may* also choose to act as mere imitators. In a sense, this points to a certain lack of robustness in the conclusion of Theorem 22, which is thus seen to depend crucially on the fact that *all* agents must *always* behave as static optimizers.

On the other hand, Theorem 25 points to the fact that such "good news" may collapse altogether if the ladder of possible sophistication is enlarged one further rung to include dynamic optimizers. In this case, the process displays the equilibrium volatility established by Theorem 23 under the assumption that *all* agents are dynamic optimizers. Thus, in line with the previous comment, equilibrium volatility is seen to be a robust phenomenon under dynamic expectations, in the sense of arising even when players may choose to behave with lower levels of sophistication.

As is apparent from the arguments used above to prove these results, the crucial fact underlying both of them may be outlined as follows. When both populations are monomorphically playing some particular equilibrium, any of the admissible types obtains the same pay-off. Therefore, evolutionary drift can "costlessly" operate on the type component of the prevailing state, leading the system to the configuration where any given contemplated transition is easiest. Borrowing a phrase from Stahl (1993), such a process of drift reflects the fact that, from an evolutionary perspective, "being right is just as good as being smart".[25],[26] Although, with a reciprocal emphasis, one could also argue that,

---

[25] Stahl (1993) analyses a context in which evolution is modelled by the Replicator Dynamics and players of different levels of "smartness" coexist. Proceeding iteratively for any level of smartness, he inductively defines the following chain. Smart$_0$ players are those who simply play any given strategy. For any $n \geq 1$, smart$_n$ players are those who choose their strategies to be a best response, given precise information on the strategy profile chosen by all less smart players and *some* probability over the strategies adopted by other players (of smartness equal to or higher than oneself) that does not violate smartness of order $n - 1$. Stahl shows that, under a variety of alternative conditions, some fraction of smart$_0$ players will survive in the long run.

[26] A related work is Banerjee and Weibull (1991). They consider a model (also based in the Replicator Dynamics) where some players of the (unique) sophisticated type are able to anticipate the strategy played by *every* one of the non-sophisticated agents they meet, while playing some rationalizable strategy among themselves. Non-sophisticated players simply adhere to one given

from an evolutionary perspective, being smart *just* involves doing what is right. "Evolution" (or, more precisely, our single-minded players) does not have, after all, any standard of smartness different from that given by pay-offs.

The previous discussion raises the question, also partially addressed by Stahl (1993), of how the analysis would be affected if sophistication is not only assumed unimportant *per se* but, quite naturally, is postulated costly. In this case (even if these costs are lexicographically less important than other pay-offs), drift within a certain "equilibrium component" can only proceed downwards (i.e. to lower levels of sophistication).[27] This, in turn, will lead to a situation where, even though higher sophistication levels might be possible, the lowest one must in the end prevail. Along the lines of Theorem 24, it is not difficult to see that such considerations restore efficiency as the long-run criterion of equilibrium selection.

strategy in all of their encounters. They are, therefore, like Stahl's (1993) smart$_0$ players. In this context, they show that the fraction of sophisticated players need not converge to 1 for some games. This happens, for example, when there is an "aggressive" strategy (like $H$ in the Hawk–Dove game of Subsection 2.3.1) whose pay-off is larger when meeting a best response than the pay-off obtained by the best responder himself.

[27] In this case, the adjustment toward lower-sophistication types would unfold (without any further mutation) as long as one mutation introduces a single representative of it. In this sense, the term "drift" applied to this phenomenon is not in line with our previous usage of it. It is still employed here, however, in order to stress the fact that the costs of sophistication could be lexicographically less important than the pay-offs derived from playing the game.

# Afterword

Having reached this point, the reader may still harbour some doubts concerning the potential of Evolutionary Theory for the analysis of social and economic situations. Indeed, such a cautionary response is fully justified, given the infant nature of the discipline, still defining its boundaries and shaping its agenda. Since, as mentioned in the Preface, this effort is very much "work in progress", it would be futile to attempt any prediction of where it will lead. Instead, let me simply conclude by suggesting three different avenues of research which, in my opinion, should be of leading concern in the near future.

- Throughout this book, the pattern in which the population (or populations) interact has been taken as given. In most cases, it was assumed to involve global random matching, although a few exceptions were also considered (e.g. local matching or "playing-the-field" contexts). An important issue which all of these models brushed aside concerns the (endogenous) evolution of the pattern of interaction itself. In many interesting cases (for example, if one is to understand the rise and evolution of new markets) a suitable account of how the network of interaction unfolds seems crucial for predicting (or influencing) the future course of events.

- The last two chapters have focused on the analysis of evolutionary systems which are perturbed by some stochastic process of mutation. In non-biological contexts, this process of mutation should be interpreted as reflecting genuinely social considerations, such as experimentation, population renewal, or some other kind of exogenous phenomenon. A more precise, well-founded formulation of evolutionary noise should significantly enrich evolutionary models in at least two respects. First, it would allow the particular specification of its crucial stochastic component to be tailored to some underlying features of the environment. Second, it would substantially help in the task of interpreting the nature and conceptual implications of its long-run predictions.

- I made the point above that, by postulating a *given* pattern of interaction, one prevents evolutionary models from analysing the important issue of how these patterns endogenously evolve. In this vein, an analogous issue may be raised in connection with the assumption that the strategy set of the game remains fixed. This, in particular, makes it impossible to analyse how genuine "mutation" (which in social environments could be interpreted as *innovation*) might affect, and in turn be affected by, evolutionary dynamics.

In a theoretical framework where new strategies may come about through some appropriately formulated process of "strategy innovation", the important issue of growth or/and technological change can be studied from an evolutionary perspective. Since these phenomena appear to be inherently evolutionary in the real world (in the sense of involving gradual, trial-and-error adjustment rather than the solution of some well-defined intertemporal optimization problem), the potential for this approach seems quite substantial.

# 7
# Appendix

## 7.1 Liapunov's Theorem

The following Theorem, due to Liapunov, represents a standard tool to tackle the issue of stability of a dynamical system. As formulated, it addresses the question of global stability. However, by a suitable application of it in a neighbourhood of a certain equilibrium, it can also be used (as in Chapter 3) for deriving conclusions of local stability.

**Theorem 26** *Let $\dot{x}(t) = H(x(t))$ be a dynamical system defined on a certain compact subset $C \subseteq \Re^n$. Let $V : C \to \Re$ be a continuously differentiable function with a unique maximum at $x^* \in C$. Then, if along any given trajectory $x(\cdot)$ of the system, $x(t) \neq x^* \Rightarrow \dot{V}(t) = D_x V(x(t)) \cdot \dot{x}(t) > 0$, then $\lim_{t \to \infty} x(t) = x^*$.*

## 7.2 Liouville's Theorem

The following result by Liouville on the conservation of volume of a dynamical system is used in the proof of Theorem 12.

**Theorem 27 (Liouville's Theorem)** *Let $\dot{x}(t) = H(x(t))$ be a dynamical system defined on a certain open subset $U \subseteq \Re^n$. Then, if $A \subseteq U$ has a volume $V \equiv \int_A dx$, then the volume $V(t)$ of the set $A(t) = \{y = x(t) : x(0) \in A\}$ satisfies:*

$$\dot{V}(t) = \int_{A(t)} div\, H(x)\, dx,$$

*where the divergence of the vector field* $H(\cdot)$ *is defined as follows:*

$$div\ H(x) \equiv \sum_{i=1}^{n} \frac{\partial H_i(x)}{\partial x_i},$$

*i.e. the trace of the Jacobian of* $H(\cdot)$.

## 7.3    A Characterization of Negative-Definiteness

The following Lemma characterizing negative definiteness of a certain matrix was required in the proof of Theorem 5, due to Hines (1980*b*).

**Lemma 29** *Let* $M$ *be any square matrix of dimension* $m$. *The following two statements are equivalent.*
*(i) For every positive definite and symmetric square matrix* $Q, QM$ *has all its eigenvalues with negative real parts;*
*(ii)* $(M + M^{\top})$ *is negative definite.*

**Proof.**
  (ii) $\Rightarrow$ (i): Let $a + ib$ be an eigenvalue of $QM$. Then, as can be seen from expressing $QM$ in Jordan canonical form, there exists at least one corresponding eigenvector $z = x + iy$ such that $x \neq 0 \neq y$ and

$$QM\,(x + iy) = (a + ib)\,(x + iy).$$

(If $b = 0$, take $y = x$.) Hence,

$$QMx = ax - by$$

and

$$QMy = bx + ay$$

which implies

$$x \cdot Mx = ax \cdot Q^{-1}x - bx \cdot Q^{-1}y$$

and

$$y \cdot My = by \cdot Q^{-1}x + ay \cdot Q^{-1}y.$$

Since $x \cdot Mx$ and $y \cdot My$ are negative, and $x \cdot Q^{-1}x$ and $y \cdot Q^{-1}y$ are positive, for $x \neq 0 \neq y$, $a$ must necessarily also be negative.
  (i) $\Rightarrow$ (ii): Since $Q$ is positive definite and symmetric, it can be expressed as $PAP^{\top}$ where $P$ consists of $m$ orthogonal column vectors $e_1, ..., e_m$ and $A$ is a diagonal matrix with positive diagonal elements $\lambda_i$.

Since all eigenvalues of $QM$ have negative real parts, the trace of $QM$ is negative. That is

$$
\begin{aligned}
tr\,(QM) &= tr\left(PAP^\top M\right) \\
&= tr\left(AP^\top MP\right) \\
&= \sum_{i=1}^{m} \lambda_i\, e_i \cdot Me_i < 0.
\end{aligned}
$$

Note that $Q$ is arbitrary so that the previous expression holds for all $P$ and $A$. Choose $x$ to be an arbitrary vector in $\Re^m$, $x \neq 0$, set $e_1 = x\,/\|x\|$, take $e_2, ..., e_m$ to be any other orthonormal vectors, and take $\lambda_1 = 1$ and $\lambda_i = \varepsilon$, for $i > 1$. Then, (i) implies that:

$$
x \cdot Mx + \varepsilon\,\|x\|^2 \sum_{i-2}^{m} e_i \cdot Me_i < 0
$$

or in the limit, as $\varepsilon \to 0$, $x \cdot Mx \leq 0$.

If $x \cdot Mx = 0$ for $x \neq 0$, $0$ is an eigenvalue of $M$, and so $\det(M) = 0$. Since $I \cdot M$ is known to have no eigenvalues with real part zero, by hypothesis, the case $x \cdot Mx = 0$ can be excluded. Since $x$ is arbitrary, $2x \cdot Mx = x \cdot \left(M + M^\top\right)x < 0$ for $x \neq 0$, so that $M + M^\top$ is negative definite, completing the proof. ∎

## 7.4  Invariant Distribution: Graph Characterization

Let $(\Omega, T)$ be a finite-state Markov process, where $\Omega$ is the (finite) state space and $T : \Omega \to \Delta(\Omega)$ stands for the transition probability function. For convenience, we shall often abuse notation and represent this function in matrix form, with $T(\omega, \omega')$ standing for the probability of a transition from state $\omega$ to state $\omega'$.

**Definition 20** *The probability distribution $\mu \in \Delta(\Omega)$ is an invariant (or stationary) distribution of the process if:*

$$
\forall \omega \in \Omega, \quad \sum_{\omega' \in \Omega} \mu(\omega')\, T(\omega', \omega) = \mu(\omega).
$$

By the standard Theory of Stochastic Processes (see, for example, Karlin and Taylor (1975: Theorem 1.3, p. 35)), if the Markov process $(\Omega, T)$ satisfies $\forall \omega, \omega' \in \Omega$, $T(\omega, \omega') > 0$, then it has a unique invariant distribution. Freidlin and Wentzel (1984) characterize this distribution as follows:

**Definition 21** *Let $\omega \in \Omega$. An $\omega$-tree $Y$ is a directed graph on $\Omega$ such that:*
*(i) Every state $\omega' \in \Omega \setminus \{\omega\}$ is the initial point of exactly one arrow; and*
*(ii) for every state $\omega' \in \Omega \setminus \{\omega\}$, there is a path linking $\omega'$ to $\omega$, i.e. there is*
*a sequence of arrows $y = \{(\omega^{(0)}, \omega^{(1)}), (\omega^{(1)}, \omega^{(2)}), ..., (\omega^{(n-1)}, \omega^{(n)})\}$ such*
*that $\omega^{(0)} = \omega'$ and $\omega^{(n)} = \omega$.*

Denoting by $\mathcal{Y}_\omega$ the set of all $\omega$ -trees, define the vector $q = (q_\omega)_{\omega \in \Omega} \in \mathfrak{R}^{|\Omega|}$
as follows:

$$q_\omega = \sum_{Y \in \mathcal{Y}_\omega} \prod_{(\omega', \omega'') \in Y} T(\omega', \omega''). \tag{7.1}$$

The following result – a particularization of Lemma 3.1 in Freidlin and
Wentzel (1984: p 177) – provides a very useful characterization of the stationary
distribution of $(\Omega, T)$ :

**Proposition 17** *The stationary distribution of $(\Omega, T)$ is given by*

$$\mu(\omega) = \frac{q_\omega}{\sum_{\omega' \in \Omega} q_{\omega'}}$$

*for all $\omega \in \Omega$.*

# Bibliography

Axelrod, R. (1984): *The Evolution of Cooperation,* New York, Basic Publ.

Axelrod, R., and W.D. Hamilton (1981): "The evolution of cooperation", *Science* **211**, 1390–6.

Auman, R., and S. Sorin (1989): "Cooperation and bounded recall", *Games and Economic Behavior* **1**, 5–39.

Banerjee, A., and J. Weibull (1991): "Evolutionary selection and rational behavior", in A. Kirman and M. Salmon, (eds.), *Rationality and Learning in Economics,* Blackwell.

Bergin, J., and B. Lipman (1995): "Evolution with state–dependent mutations", *Econometrica,* forthcoming.

Bernheim, D. (1984): "Rationalizable strategic behavior", *Econometrica* **52**, 1007–28.

Bhaskar, V. (1994): "Noisy communication and the fast evolution of cooperation", CentER Discussion Paper no. 94112.

Binmore, K., and L. Samuelson (1992): "Evolutionary stability in games played by finite automata", *Journal of Economic Theory* **57**, 278–305.

Binmore, K., L. Samuelson, and R. Vaughan (1993): "Musical chairs: modeling noisy evolution", mimeo, University College (London) and University of Wisconsin.

Björnerstedt, J., M. Dufwenberg, P. Norman and J. Weibull (1993): "Evolutionary selection dynamics and irrational survivors", mimeo, Dept. of Economics, Stockholm University.

Bomze, I. M. (1986): "Non–Cooperative, Two–Person Games in Biology: A Classification", *International Journal of Game Theory* **15**, 31–57.

Bomze, I. M., and B. M. Pötscher (1988): *Game Theoretic Foundations of Evolutionary Stability,* Berlin, Springer–Verlag.

Bowler, P. J. (1984): *Evolution: The History of an Idea,* University of California Press.

Boyd, R., and P. Richerson (1985): *Culture and the Evolutionary Process,* Chicago, University of Chicago Press.

Brandemburger, A., and E. Dekel (1987): "Rationalizability and correlated equilibrium", *Econometrica* **55**, 1391–402.

Cabrales, A. (1992): "Stochastic replicator dynamics", mimeo, University of California at San Diego.

Cabrales, A., and J. Sobel (1992): "On the limit points of discrete selection dynamics", *Journal of Economic Theory* **57**, 407–19.

Crawford, V. (1991): "An 'evolutionary' interpretation of van Huyck, Battalio, and Beil's experimental results on coordination", *Games and Economic Behavior* **3**, 25–59.

Dawkins, R. (1982): *The Selfish Gene,* Oxford and San Francisco, Freeman.

Dekel, E., and S. Scotchmer (1992): "On the evolution of optimizing behavior", *Journal of Economic Theory* 57, 392–406.

Diamond, P. A. (1982): "Aggregate–demand management in search equilibrium", *Journal of Political Economy* 4, 881–94.

Ellison, G. (1993): "Learning, local interaction, and coordination", *Econometrica* 4, 1047–73.

Eshel, I. (1982): 'Evolutionary stable strategies and viability selection in Mendelian populations", *Theoretical Population Biology* 22, 204–17.

Eshel, I. (1991): "Game Theory and population dynamics in complex genetical systems: the role of sex in short term and long term evolution", in R. Selten (ed.), *Game Equilibrium Models, vol. I, Evolution and Game Dynamics*, Berlin, Springer-Verlag, 6–28.

Eshel, I., and E. Akin (1983): "Co–evolutionary instability of mixed Nash solutions", *Journal of Mathematical Biology* 18, 123–33.

Fogel, D. (1996): "Special Issue on the Prisoner's Dilemma", *BioSystems* 37.

Foster, D., and P. Young (1990): "Stochastic evolutionary game dynamics", *Theoretical Population Biology* 38, 219–32.

Freidlin, M. I., and A. D. Wentzel (1984): *Random Perturbations of Dynamical Systems*, New York, Springer-Verlag.

Friedman, D. (1991): "Evolutionary games in Economics", *Econometrica* 59, 637–66.

Friedman, J. (1977): *Oligopoly and the Theory of Games*, Amsterdam, North Holland.

Fudenberg, D., and C. Harris (1992): "Evolutionary dynamics with aggregate shocks", *Journal of Economic Theory* 57, 420–41.

Fudenberg, D., and E. Maskin (1990): "Evolution and cooperation in noisy repeated games", *American Economic Review* 80, 274–9.

Fudenberg, D., and J. Tirole (1991): *Game Theory*, Cambridge, Mass., MIT Press.

Gale, J., K. Binmore, and L. Samuelson (1995): "Learning to be imperfect: the ultimatum game", *Games and Economic Behavior* 8, 56–90.

Gihman, I., and A. V. Skorohod (1972): *Stochastic Differential Equations*, Berlin, Springer-Verlag.

Güth, W., R. Schmittberger, and B. Schwarze (1982): "An experimental analysis of ultimatum bargaining", *Journal of Economic Behavior and Organization* 3, 367–88.

Hamilton, W. D. (1970): "Selfish and spiteful behavior in an evolutionary model", *Nature* 228, 1218–20.

Hammerstein, P., and G. A. Parker (1981): "The asymmetric war of attrition", *Journal of Theoretical Biology* 96, 647–82.

Hammerstein, P., and R. Selten (1992): "Evolutionary Game Theory", *Handbook of Game Theory*, in R. Aumann and S. Hart (eds.), Amsterdam, North–Holland.

Harsanyi, J. C., and R. Selten (1988): *A General Theory of Equilibrium Selection*, Cambridge, Mass., MIT Press.

Hildenbrand, W. (1974): *Core and Equilibria of a Large Economy*, Princeton, Princeton Univ. Press.

Hines, W. G. S. (1980*a*): "Three characterizations of population strategy stability", *Journal of Applied Probability* **17**, 333–40.

Hines, W.G.S. (1980*b*): "Strategy stability in complex populations", *Journal of Applied Probability* **17**, 600–10.

Hirsch, M. W., and S. Smale (1974): *Differential Equations, Dynamical Systems and Linear Algebra*, New York, Academic Press.

Hirshleifer, J., and J. C. Martinez Coll (1992): "Selection, mutation and the preservation of diversity in evolutionary games", *Revista Española de Economía* **9**, 251–73.

Hofbauer, J., P. Schuster, and K. Sigmund (1979): A note on evolutionary stable strategies and game dynamics", *Journal of Theoretical Biology* **81**, 609–12.

Hofbauer, J., and K. Sigmund (1987): "Permanence for replicator equations" in A. Kurzhanski and K. Sigmund (eds.), *Dynamical Systems*, Lecture Notes in Economics and Mathematical Systems 287, Berlin, Springer-Verlag.

Hofbauer, J., and K. Sigmund (1988): *Dynamical Systems and the Theory of Evolution*, Cambridge, Cambridge University Press.

Hutson, V., and W. Moran (1982): "Persistence of species obeying difference equations", *Mathematical Biosciences* **63**, 253–69.

Kandori, M., G. Mailath, and R. Rob (1993): "Learning, mutation, and long–run equilibria in games", *Econometrica* **61**, 29–56.

Kandori, M., and R. Rob (1995): "Evolution of equilibria in the long run: a general theory and applications", *Journal of Economic Theory* **65**, 383–414.

Karlin, S., and H.M. Taylor (1975): *A First Course in Stochastic Processes*, London, Academic Press.

Kim, Y.–G., and J. Sobel (1995): "An evolutionary approach to pre–play communication", *Econometrica* **63**, 1181–93.

Kimura, M. (1983): *The Neutral Theory of Molecular Evolution*, Cambridge, Cambridge University Press.

Kohlberg, E., and J.–F. Mertens (1986): "On the strategic stability of equilibria", *Econometrica* **54**, 1003–37.

Losert, V., and E. Akin (1983): "Dynamics of games and genes: discrete versus continuous time", *Journal of Mathematical Biology* **17**, 241–51.

Mailath, G. (1992): "Introduction: Symposium on Evolutionary Game Theory", *Journal of Economic Theory* **57**, 259–77.

Mas–Colell, A. (1980): "Non–cooperative approaches to the theory of perfect competition", *Journal of Economic Theory* **22**, 121–376.

Matsui, A. (1991): "Cheap talk and cooperation in a society", *Journal of Economic Theory* **54**, 245–58.

Maynard Smith, J. (1982): *Evolution and the Theory of Games*, Cambridge, Cambridge University Press.

Maynard Smith, J., and G. Price (1973): "The logic of animal conflicts", *Nature* **246**, 15–18.

Milgrom, P., and J. Roberts (1991): "Adaptative and sophisticated learning in normal form games", *Games and Economic Behavior* **3**, 82–100.

Nachbar, J. H. (1990): "Evolutionary selection in dynamic games", *International Journal of Game Theory* **19**, 59–90.

Nash, J. (1950): "The bargaining problem", *Econometrica* **18**, 155–62.

Nöldeke, G., and L. Samuelson (1993): "An evolutionary analysis of backward and forward induction", *Games and Economic Behavior* **5**, 425–55.

Nowak, M., and K. Sigmund (1992): "Tit for Tat in heterogeneous populations", *Nature* **355**, 250–2.

Oechssler, J. (1993): "Competition among conventions", mimeo, Columbia University.

Pearce, D. (1984): "Rationalizable strategic behavior and the problem of perfection", *Econometrica* **52**, 1029–50.

Peleg, B., and A. Shmida (1992): "Short–run stable matchings between bees and flowers", *Games and Economic Behavior* **4**, 232–51.

Primack, A. (1985): "Patterns of flowering phenology in communities, populations, individuals, and single flowers", in J. White (ed.), *The Population Structure of Vegetation*, Dordrecht, The Netherlands, Junk, 571–593.

Rhode, P., and M. Stegeman (1995): "Non–Nash equilibria of Darwinian dynamics", mimeo.

Rhode, P., and M. Stegeman (1996): "A comment on learning, evolution, and long–run equilibria in games", *Econometrica* **64**, 443-50.

Ritzberger, K., and K. Vogelsberg (1990): "The Nash field", IAS research report no. 263, Vienna.

Ritzberger, K., and J. Weibull (1995): "Evolutionary selection in normal–form games", *Econometrica* **63**, 1371-400.

Robson, A. (1990): "Efficiency in evolutionary games: Darwin, Nash, and the secret handshake", *Journal of Theoretical Biology* **144**, 379–96.

Robson, A. (1993): "The 'Adam and Eve effect' and fast evolution of efficient equilibria", mimeo, University of Western Ontario.

Robson, A., and F. Vega–Redondo (1996): "Efficient equilibrium selection in evolutionary games with random matching", *Journal of Economic Theory*, forthcoming.

Samuelson, L. (1988): "Evolutionary foundations of solution concepts for finite, two–player, normal–form games", in M.Y. Vardi (ed.), *Theoretical Aspects of Reasoning About Knowledge,* Los Altos, Morgan Kauffman, 221–5.

Samuelson, L. (1990): "Limit evolutionarily stable strategies in asymmetric games", *Games and Economic Behavior* **3**, 110–28.

Samuelson, L. (1994): "Stochastic stability in games with alternative best replies", *Journal of Economic Theory* **64**, 35–65.

Samuelson, L., and J. Zhang (1992): "Evolutionary stability in asymmetric games", *Journal of Economic Theory* **57**, 363–91.

Schaffer, M. E. (1988): "Evolutionary stable strategies for a finite population and a variable contest size", *Journal of Theoretical Biology* **132**, 469–78.

Schaffer, M.E. (1989): "Are profit maximisers the best survivors?: A Darwinian model of economic natural selection", *Journal of Economic Behavior and Organization* **12**, 29–45.

Selten, R. (1980): "A note on evolutionary stable strategies in asymmetric animal conflics", *Journal of Theoretical Biology* **84**, 93–101.

Selten, R. (1983): "Evolutionary stability in extensive two–person games", *Mathematical Social Sciences* **5**, 269–363.

Selten, R. (1988): "Evolutionary stability in extensive two–person games: correction and further development", *Mathematical Social Sciences* **16**, 223–66.

Selten, R., and A. Shmida (1991): "Pollinator foraging and flower competition in a game equilibrium model", in R. Selten (ed.), *Game Equilibrium Models I: Evolution and Game Dynamics,* Berlin, Springer-Verlag.

Sigmund, K. (1993): *Games of Life: Explorations in Ecology, Evolution, and Behaviour,* Oxford, Oxford University Press.

Smallwood, D., and J. Conlisk (1979): "Product quality in markets where consumers are imperfectly informed", *Quarterly Journal of Economics* **93**, 1–23.

Sobel, J. (1993): "Evolutionary stability and efficiency", *Economic Letters* **42**, 313–19.

Stahl, D.O. (1993): "Evolution of smart$_n$ players", *Games and Economic Behavior* **5**, 604–17.

Swinkels, J. (1992): "Evolutionary stability with equilibrium entrants", *Journal of Economic Theory* **57**, 306–32.

Swinkels, J. (1993): "Adjustment dynamics and rational play in games", *Games and Economic Behavior* **5**, 455–84.

Taylor, P. D., and L. B. Jonker (1978): "Evolutionary stable strategies and game dynamics", *Mathematical Bioscience* **40**, 145–56.

Thomas, B. (1985): "On evolutionary stable sets", *Journal of Mathematical Biology* **22**, 105–15.

Topkis, D. (1979): "Equilibrium points in nonzero–sum $n$–person submodular games", *SIAM Journal of Control and Optimization* **17**, 773–87.

Van Damme, E. (1987): *Stability and the Perfection of Nash Equilibria,* Berlin, Springer-Verlag.

Van Damme, E. (1989): "Stable equilibria and forward induction", *Journal of Economic Theory* **48**, 476–96.

Vega–Redondo, F. (1993): "Competition and culture in the evolution of economic behavior", *Games and Economic Behavior* **5**, 618–31.

Vega–Redondo, F. (1995): "Expectations, drift, and volatility in evolutionary games", *Games and Economic Behavior* **11**, 391–412.

Vega–Redondo, F. (1996*a*): "Long–run cooperation in the one–shot Prisoner's Dilemma", *BioSystems* **37**, 39–47.

Vega–Redondo, F. (1996*b*): "Pollination and reward: a game–theoretic approach", *Games and Economic Behavior* **12**, 127–43.

Vega–Redondo, F. (1996*c*): "The evolution of Walrasian behavior", *Econometrica,* forthcoming.

Wärneryd, K. (1993): "Cheap talk, coordination, and evolutionary stability", *Games and Economic Behavior* **5**, 532–46.

Wright, S. (1945): "Tempo and mode in evolution", *Ecology* **26**, 415–19.

Young, P., and D. Foster (1991): "Cooperation in the short and in the long run", *Games and Economic Behavior* **3**, 145–56.

Young, P. (1993*a*): "The evolution of conventions", *Econometrica* **61**, 57–84.

Young, P. (1993*b*): "An evolutionary model of bargaining", *Journal of Economic Theory* **59**, 145–68.

Zeeman, E.C. (1981): "Dynamics of the evolution of animal conflicts", *Journal of Theoretical Biology* **89**, 249–70.

# Index